IMAGES of America
WEST TEXAS CATTLE KINGDOM

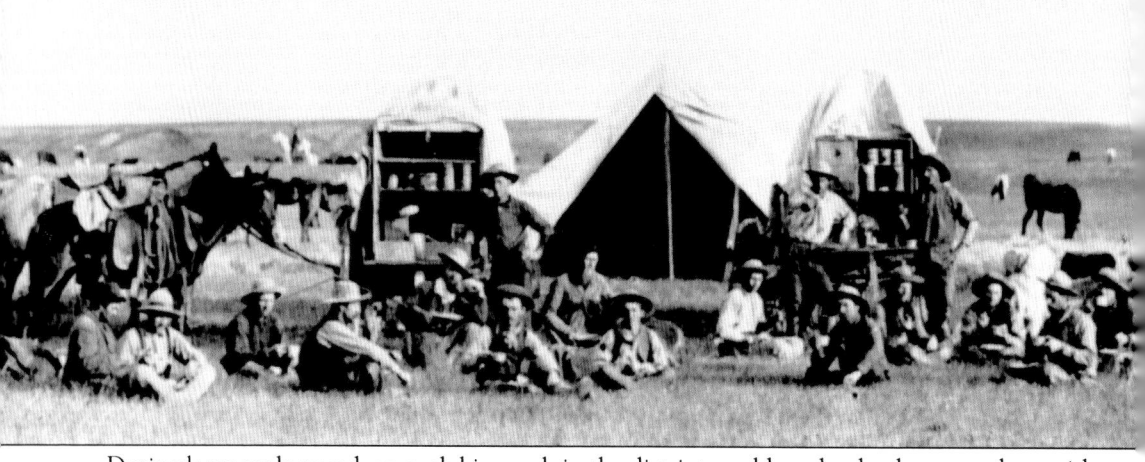

During large-scale roundups, each big ranch in the district would send a chuck wagon along with its cowboys. The combined crews were then well fed by a number of cooks and chuck wagons. Here, a tent has been pitched between two chuck wagons at this roundup camp. (Courtesy Panola College Library, Carthage, Texas.)

ON THE COVER: Cowboys from the vast JAs Ranch pull up long enough to create an interesting water reflection. Note the large lariats on their saddles. (Courtesy Panhandle-Plains Museum, Canyon, Texas.)

IMAGES
of America

WEST TEXAS CATTLE KINGDOM

Bill O'Neal

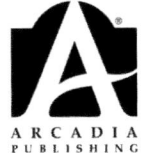

Copyright © 2013 by Bill O'Neal
ISBN 978-0-7385-9648-8

Published by Arcadia Publishing
Charleston, South Carolina

Printed in the United States of America

Library of Congress Control Number: 2012947933

For all general information, please contact Arcadia Publishing:
Telephone 843-853-2070
Fax 843-853-0044
E-mail sales@arcadiapublishing.com
For customer service and orders:
Toll-Free 1-888-313-2665

Visit us on the Internet at www.arcadiapublishing.com

For my great-grandfather, trail driver Jess Standard; my grandfather, cowboy Will Standard; and my uncle, rodeo cowboy Ted Standard

Contents

Acknowledgments		6
Introduction		7
1.	Origins	9
2.	Texas Cattle Trails	17
3.	Charles Goodnight	33
4.	Great Ranches and Ranchers	43
5.	Cowboys—and Cowgirls	67
6.	Texas Cattle Towns	87
7.	Trouble on the Range	97
8.	From Rodeos to Reel Cowboys	111
9.	Where to Go and What to See	119

Acknowledgments

Several months ago, Kristie Kelly, publisher for Arcadia Publishing, discussed with me the possibilities of creating a book about Texas cowboys, ranches, longhorn cattle, trail drives, and range troubles. I have written and lectured about this classic story for decades and was immediately enthusiastic about the possibilities of relating this frontier saga through the Arcadia treatment. I am grateful to Kristie for including me in this project and to my editor, Laura Bruns, for her courteous and efficient assistance throughout the development of Images of America: *West Texas Cattle Kingdom*.

Having written a number of previous books and magazine articles on this subject, I already possessed a large number of images that could be used in the book. Nonetheless, there was a need for many more photographs, and late in the spring of 2012, my wife and I took two trips to West Texas. At the Scurry County Museum in Snyder, we were given major assistance by director Daniel Schlegel, curator Sarah Bellian, and administrative assistant Janie Guerrero. Rich photographic files were opened to us by Sarah Bellian, who also led us through the museum's extensive storage holdings so that we could photograph artifacts not on exhibit.

The Charles Goodnight home at Goodnight has recently undergone extensive renovations to restore its original appearance. When we stopped by for photography, we were approached by the genial Emery Gooden. Emery and his wife, Montie, are key members of the Armstrong County Historical Society and were instrumental in the restoration effort. Emery revealed to us a great deal about the Goodnight house and the restoration project.

Ruth Crawford presides over the Bosque County Collection, located in the county museum on the courthouse square in Meridian. Ruth graciously made available to me a number of previously unpublished photographs. I am likewise grateful to Carol Northington Wright of Lampasas for sharing with me images and information about her great-grandfather Alex Northington, a frontier rancher and state legislator. Betty M. Giddens, a descendant of a prominent ranching family and of rancher/trail boss/feudist Pink Higgins, also provided photographs of her pioneer ancestors.

On two visits to the Coryell County Museum in Gatesville, I was assisted by several cordial and informative docents. At the Panola College Library in Carthage, Sherri Baker provided her customary aid in locating materials for this project.

As always, my wife, Karon, offered invaluable help. She accompanied me on research trips, scanned photographs, and served as a sounding board. And it would not have been possible to meet the necessary deadlines without her tireless efforts at manuscript and image preparation.

Introduction

After the Civil War, the American public became captivated by the frontier world of hard-riding cowboys and wild longhorn cattle and enormous ranches. This compelling way of life originated and evolved in Texas. Both cattle and horses were introduced to the Western Hemisphere by Spanish conquistadores and colonizers. On the ranges of northern Mexico, vaqueros handled cattle from horseback, developing special techniques, tools, and attire. Roping, branding, heavy-duty saddles, jingling spurs, leather *chaparejos*, high-heeled boots, wide-brimmed sombreros—everything had utilitarian purposes but came to seem colorful and even romantic.

Through the years, countless expeditions from Mexico marched into Texas along El Camino Real (The Royal Highway), some going as far as Spanish Florida. Every expedition drove cattle, often to stock the Catholic missions in Texas. During the 1760s, Mission La Bahia maintained a herd of 15,000, Mission Rosario had 10,000 head, and most of the other two dozen Texas missions also kept substantial herds. During the Spanish journeys into or through Texas, cattle inevitably strayed into brush country north of the Rio Grande, multiplying freely in this unpopulated region. It was a harsh land where cattle had to become hardy survivors, good at finding water and any kind of forage, and aggressive in battling predators—with horns that evolved into long, dangerous weapons.

In the 1800s, Anglo-American frontiersmen in Texas adapted the techniques and equipment of the vaqueros. Texas "cow boys" drove a few herds of longhorns to distant markets such as New Orleans and California during the 1840s and 1850s. But before these long drives could become a regular activity, the Civil War blocked Texas from almost all cattle markets, while most able-bodied men served with the Confederate army.

Unattended and ignored, longhorn cattle multiplied prolifically, especially within a vast natural corral enclosed by the Rio Grande on the south and west, the Gulf of Mexico on the east, and the Nueces River across the north. By the end of the Civil War, as many as five million wild longhorns ranged across the grasslands of Texas, while a hungry market for beef opened in the industrial Northeast. Longhorns costing no more than $3 or $4 in Texas would bring $30 to $50 in northern markets. Herds of half-wild cattle were rounded up by Texas drovers and driven toward the nearest railroads, first to Missouri along the Sedalia Trail, then to Kansas along the Chisholm Trail and the Great Western Trail. Other trails radiated out of the West Texas Cattle Kingdom, most importantly the Goodnight-Loving Trail.

The frontier long had attracted adventurous youngsters, and now adolescents and young men eagerly came to Texas to become cowboys on trail drives. Booted and spurred, clad in big hats and chaps and bandanas, cowboys were high-spirited, proud, and tough.

"Timid men were not among them," admired legendary Texas cattleman Charles Goodnight. Cowboys possessed the majestic feeling of power and height and superiority of mounted men throughout history, but their work proved hard and dangerous. Cattle drives bristled with hazards, particularly treacherous river crossings and stampedes. Longhorns were quick, ornery beasts capable of inflicting harm upon men and horses. Cowboys had to ride, rope, and master other athletic skills to handle those cantankerous creatures.

The central workplace of the cowboy was the ranch, a Cattle Kingdom institution which rapidly acquired its own magnetic appeal. During the formative period of the range cattle industry, enormous ranches grew up, encompassing hundreds of thousands—sometimes millions—of acres of grasslands. The first great ranch of the Cattle Kingdom was established in the 1850s by Richard King, a tough visionary who created a world-famous empire. Within half a century, 75,000 cattle, along with 10,000 horses and mules, grazed on 1,150,000 King Ranch acres.

Charles Goodnight's most famous ranch was the JAs, centered in magnificent Palo Duro Canyon. Backed by wealthy Englishman John Adair, Goodnight acquired other nearby ranches. Soon 100,000 JA cattle grazed on more than 1,335,000 acres. The flamboyant Shanghai Pierce organized El Rancho Grande in 1865, branding as many as 18,000 calves a year. The biggest ranch of all was the XIT, over 3,000,000 acres extending from the northwest corner of the Texas Panhandle for more than 200 miles along the Texas/New Mexico border. During its heyday, the XIT maintained herds of 125,000 to 150,000 cattle.

Life on these ranches was vigorous and exhilarating. Admittedly, existence on smaller, hardscrabble spreads could be brutally monotonous and narrow, but outfits with thousands or tens of thousands of cattle radiated success and excitement. There was a festive, circus-like atmosphere at big roundups, and large crews felt a strong sense of camaraderie and loyalty to "their" ranch.

Henrietta King, known as "La Madama" by the *Kiñenos* who lived and worked on the King Ranch, was the ultimate authority for four decades after the death of her husband. Molly Goodnight mothered her husband's riders and was revered by her husband for her unfailing cheerfulness and courage in standing up to hardships and long periods of complete isolation from other women. A great many other ranch wives endured hardship and danger, and they and their daughters often saddled up and performed range work.

At least once King Ranch headquarters was attacked by outlaws, and Richard King often had to travel with an armed escort of a dozen or more riders. With rustling rampant, King and his neighbor and one-time partner, Mifflin Kennedy, founded the Stock Raisers Association of Western Texas. Rustlers and brand burners plagued Shanghai Pierce—until he led 80 men into their camp and lynched five stock thieves. The Horrell brothers of Lampasas County raided the herd of Pink Higgins, triggering the Horrell-Higgins Feud, a bloody range war. In 1877, forty Texas ranchers formed the Stock-Raisers Association of Northwest Texas to battle rustlers (today the organization is the 15,000-member Texas and Southwestern Cattle Raisers Association).

There were fence-cutting incidents and range conflicts between cattlemen and sheepherders. In Tascosa, the "Cowboy Capital of the Panhandle," there were cowboy gunfights throughout the 1880s, as well as a cowboy strike. Fort Worth and San Antonio experienced cattle town violence, including shoot-outs involving such prominent gunfighters as Like Short, Ben Thompson, King Fisher, "Longhair Jim" Courtright, and Jim "Killer" Miller, the West's premier assassin.

The West Texas Cattle Kingdom was celebrated through a variety of venues. Pecos staged the West's first rodeo. An African American cowboy, Bill Pickett, introduced bulldogging to the rodeo world. Singing cowboys Gene Autry and Tex Ritter were native Texans, while cowboy movie superstar Tom Mix, from Pennsylvania, claimed to have been born in West Texas. *Red River*, starring John Wayne, and *The Unforgiven*, with Burt Lancaster and Texan Audie Murphy, were among the best motion pictures ever filmed about Texas cowboys and ranching. For more than three decades, *Giant* was called "the national movie of Texas" before losing this informal title to *Lonesome Dove*. *Giant* was filmed in West Texas, while *Lonesome Dove* starred Tommy Lee Jones, who still works his San Saba ranch. The immensely popular television miniseries featured an epic cattle drive and climaxed with an unforgettable incident based on Charles Goodnight's journey back to Texas with the corpse of his partner, Oliver Loving. And the miniseries was based on the Pulitzer Prize-winning novel *Lonesome Dove*, written by Larry McMurtry, who was raised on a West Texas cattle ranch.

One

ORIGINS

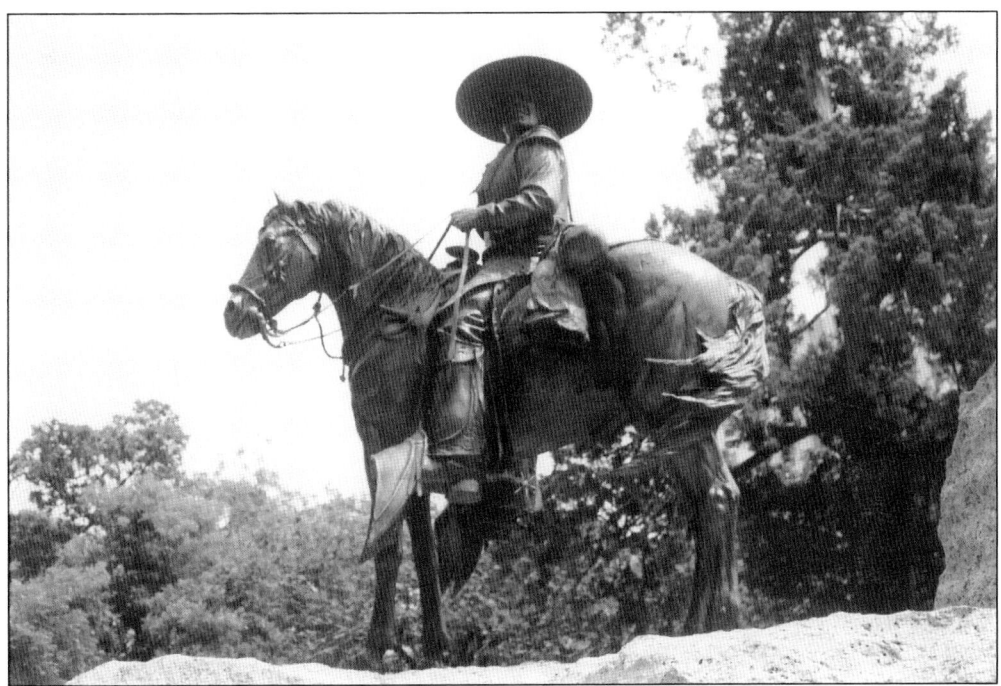

Spanish explorers and colonizers introduced cattle and horses to the New World in the 1500s. The hacienda system was developed by the Spanish, and vaqueros worked cattle from horseback on great ranchos. Proud vaqueros donned big sombreros and boots and spurs, and their attire, equipment, and techniques were adapted by Anglo "cow boys" in the 1800s. This vaquero statue is part of a Tejano group unveiled in 2012 on the grounds of the Texas State Capitol in Austin. (Photograph by the author.)

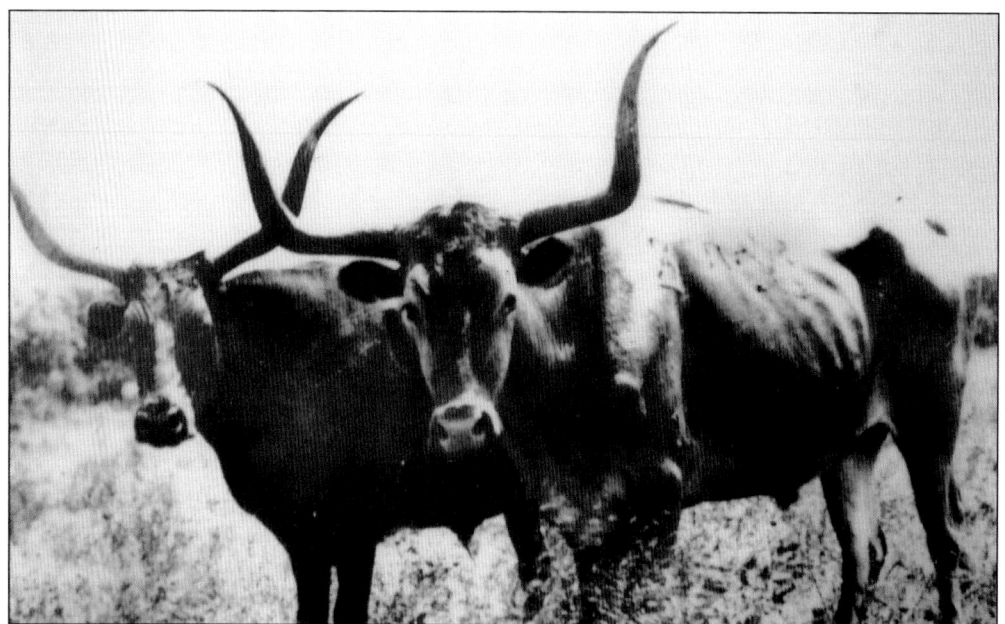

Longhorn cattle were lean, agile, and tough. In the brush country of south Texas, these half-wild beasts became hardy survivors, finding water and foraging for food. They were also aggressive when fighting off predators, with their horns evolving into long, dangerous weapons. (Courtesy Historical and Biographical Record of the Cattle Industry of Texas.)

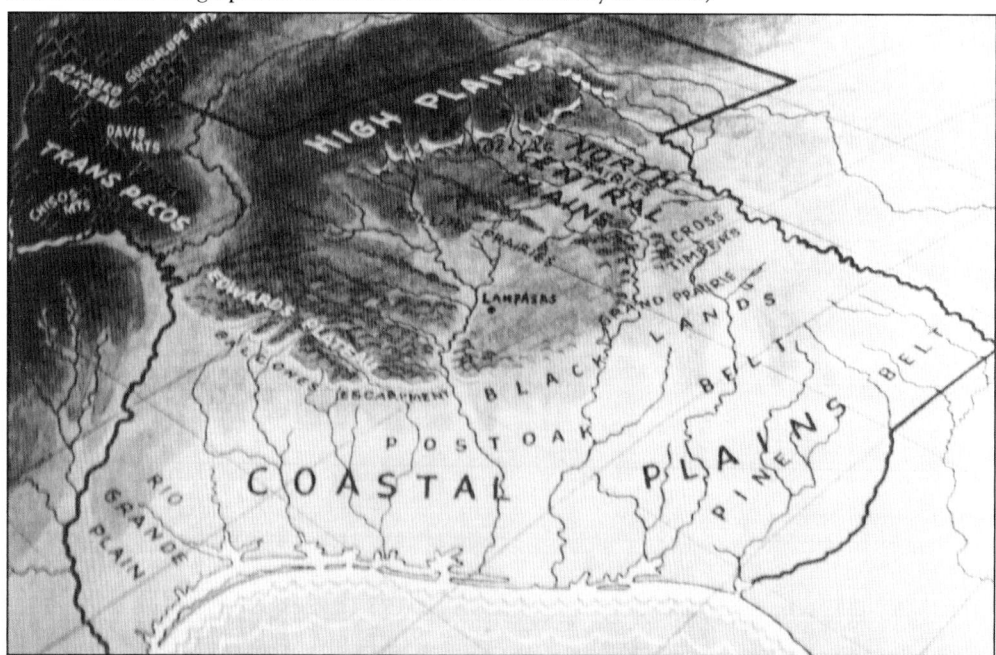

Almost all Spanish expeditions entering Texas came up El Camino Real through the Rio Grande Plain. These expeditions brought cattle, some of which strayed into the harsh brush country. Eventually, millions of longhorns bred within a vast natural corral bound by the Gulf of Mexico, the Rio Grande (at left), and the Nueces River, which curves over the "C" and between the "C" and "O" in the word "Coastal." (Courtesy Texas Almanac, 1939–1940.)

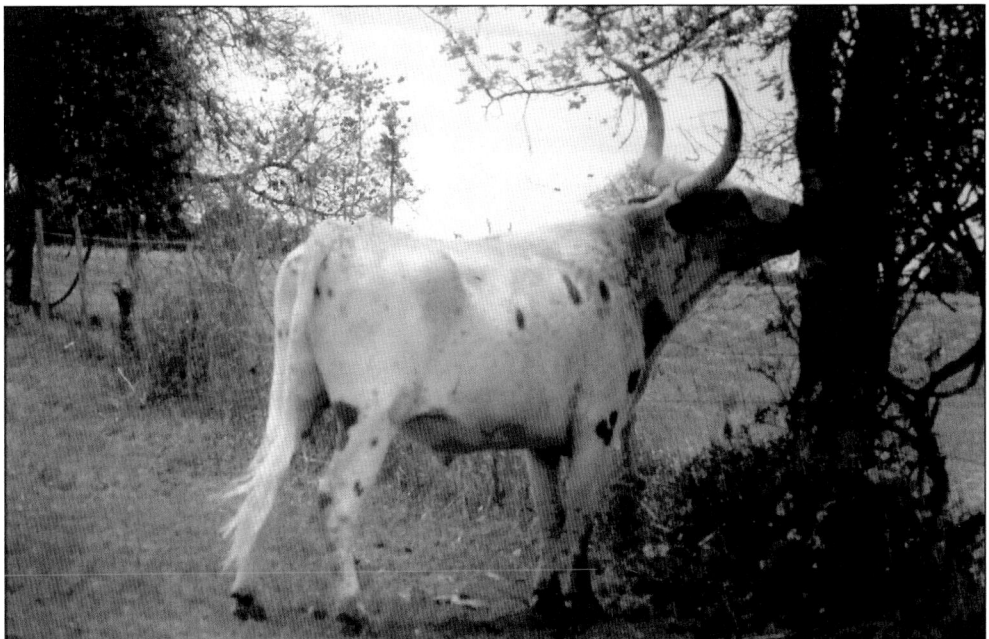

Longhorns browsed food from trees like deer, digesting mesquite beans. Although mesquite trees were originally confined to south Texas, when longhorns began to be driven north by the hundreds of thousands, elimination and the trampling of seeds into the earth caused the spread of mesquites across the state. (Photograph by the author.)

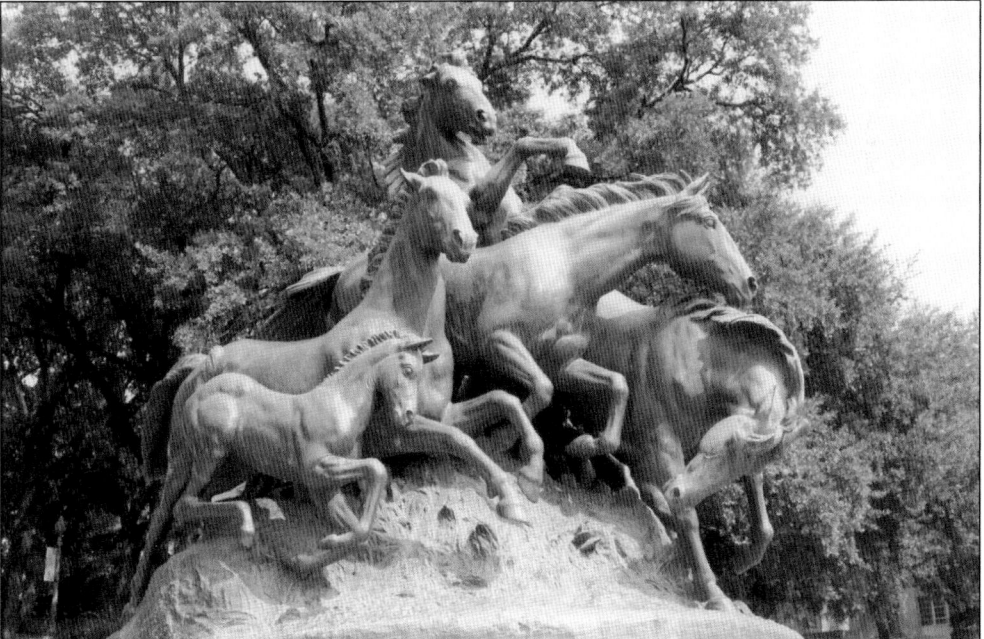

Mustangs—from the Spanish *mesteño*, or "stray"—multiplied rapidly and grew up wild in Texas and across much of the West. These small, grass-fed ponies were rounded up to be broken for use on ranchos at about the age of four. The vaquero, proud of his equestrian skills, viewed his world from four feet above the ground. (Photograph by the author.)

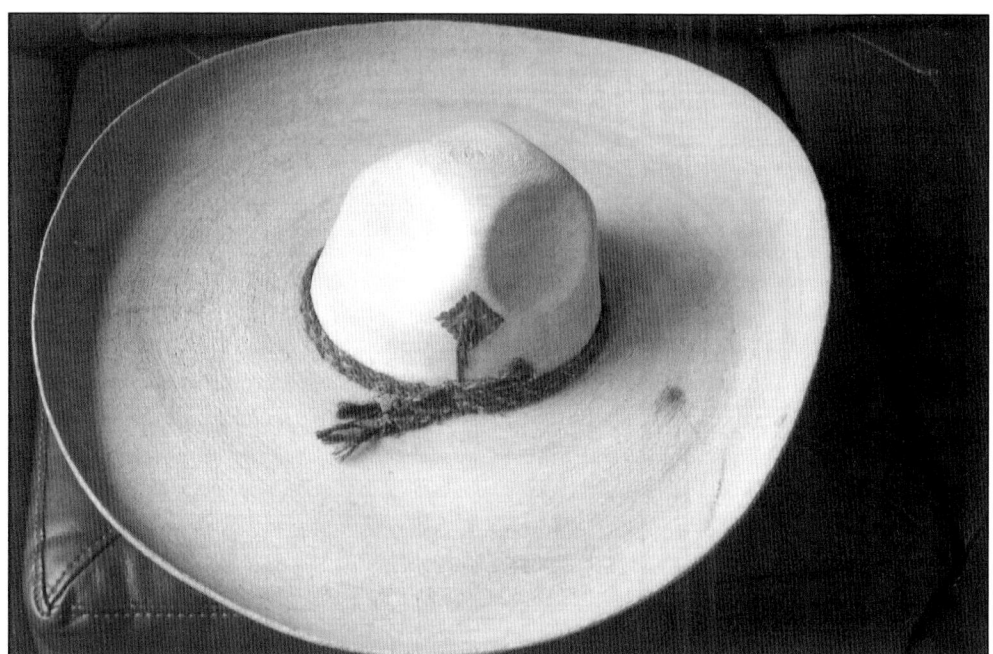

Broad-brimmed sombreros were worn by vaqueros and served many uses. As Anglo cowboys learned, the big hat offered protection from sun, rain, hail, sleet, and snow. The crown could double as a water bucket or grain bag for his horse, and if there was a chance for a little daytime sleep, the sombrero was a perfect eyeshade. (Photograph by the author.)

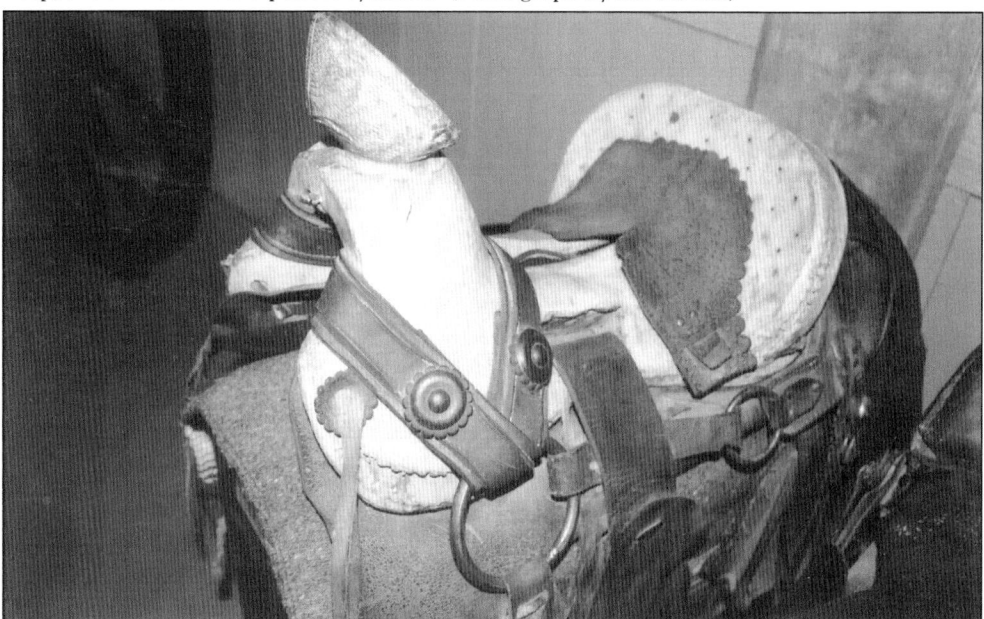

Vaqueros developed a large stock saddle based on the Spanish war saddle. The broad saddle horn in front was used to wrap the loose ends of *la reata*—"the rope," later anglicized to "lariat." *Dar la vuelta*, or "to turn over," described the movement, a phrase anglicized to "dally." Cowboys would dally the end of the lariat around the saddle horn. This saddle is kept at the Texas Cowboy Hall of Fame at the Fort Worth Stockyards. (Photograph by the author.)

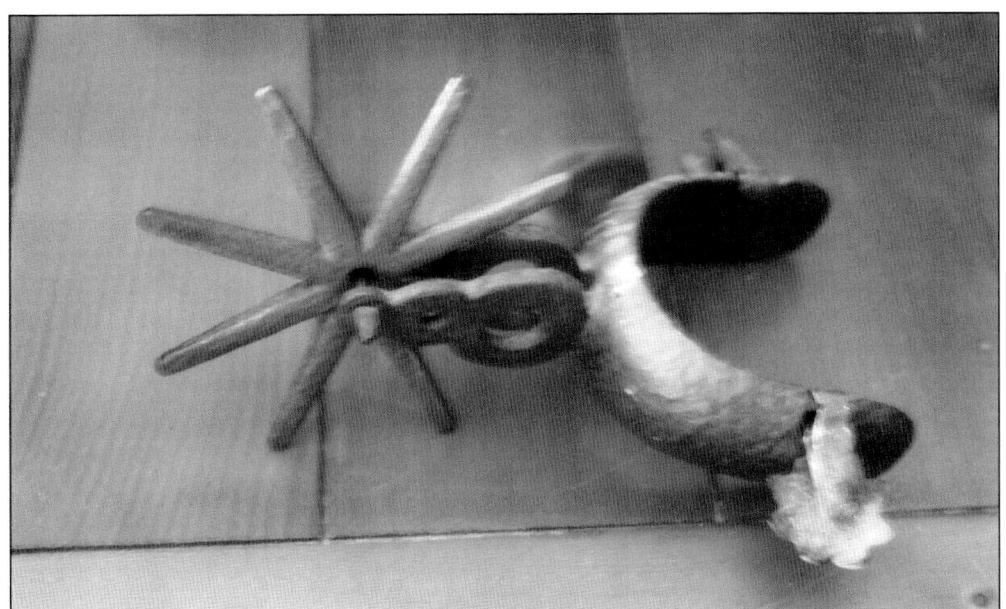

Spurs, like the ones above, are essential to controlling a horse, especially when encouraging him to cross a rough place he is reluctant to tackle or to signal him for quick turns. Vaqueros favored Chihuahua spurs, featuring rowels of six inches or more in length. Such spurs had to be buckled on after the vaquero was in the saddle. (Photograph by the author.)

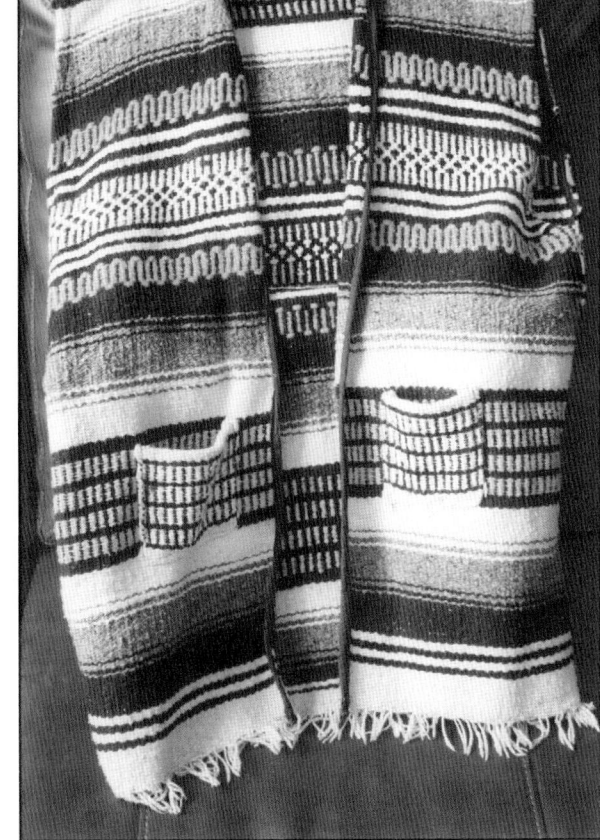

A woven vest provided warmth to vaqueros, with easy-to-reach pockets in front, while leaving his arms free to work with *la reata*. Like so many other items of attire and equipment, the vest was adapted by American cowboys, often as a canvas garment with four front pockets. (Photograph by the author.)

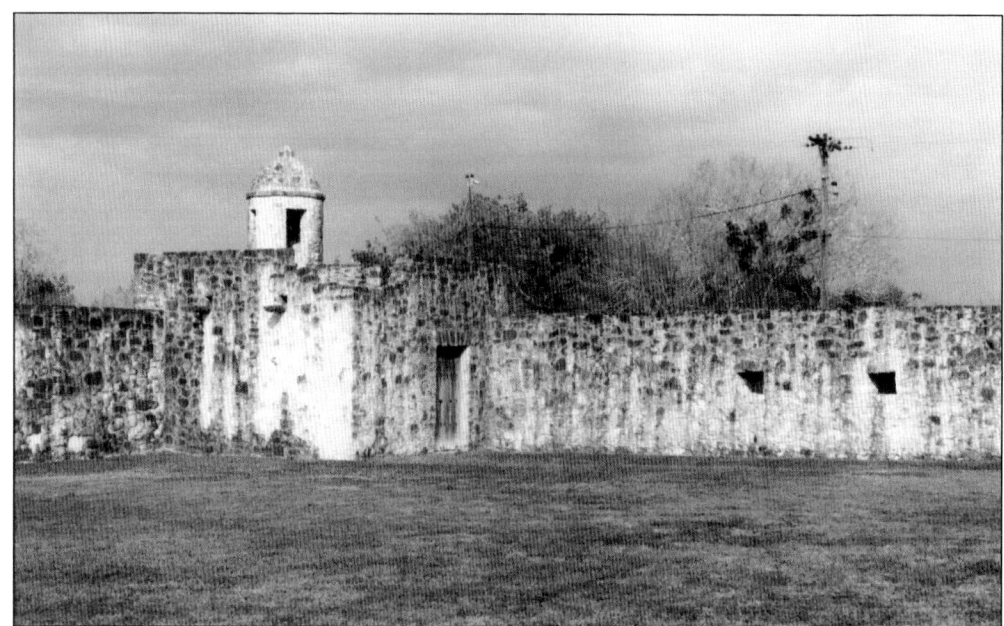

From 1690 through 1792, the Spanish established 25 missions in northern Texas. It was the purpose of the Franciscan missionaries to convert the Indians to Catholic Christianity, to make them *gente de razon*, or "people of reason," and to teach them to farm and raise livestock. Most missions established nearby cattle ranches. Mission La Bahía del Espíritu Santo at Goliad was a large complex that included an impressive presidio (above) and a ranch where 15,000 head of longhorns grazed. A few miles to the west, Mission Rosario grew a cattle herd that numbered 30,000 by 1780. The most successful mission effort was at San Antonio, where five missions were established in the 1700s. The most impressive complex was Mission San Jose (below), but all five had cattle ranches. (Photographs by the author.)

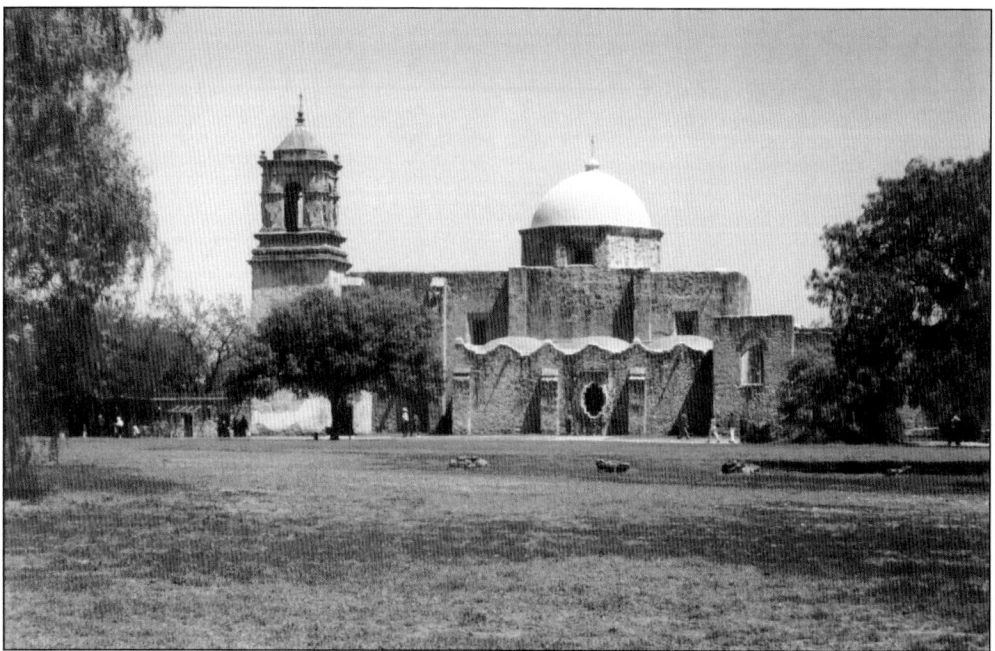

Richard King arrived in Texas in 1847 at the age of 22 as a riverboat pilot for the US Army. Following the Mexican War, King built a profitable steamboat business along the Rio Grande. Within a few years, he began ranching, carefully acquiring title to large tracts of land, while buying cattle cheaply below the border. In time, 75,000 head of cattle grazed more than one million acres of the fabled King Ranch. (Courtesy Historical and Biographical Record of the Cattle Industry of Texas.)

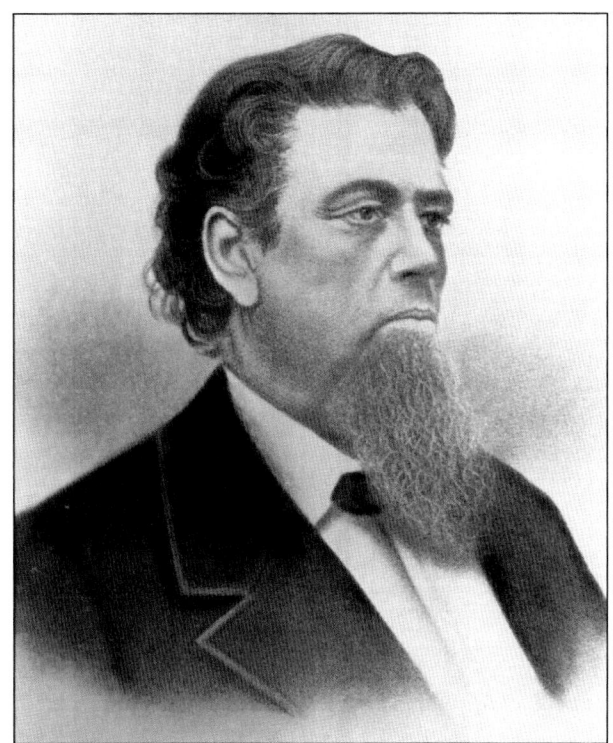

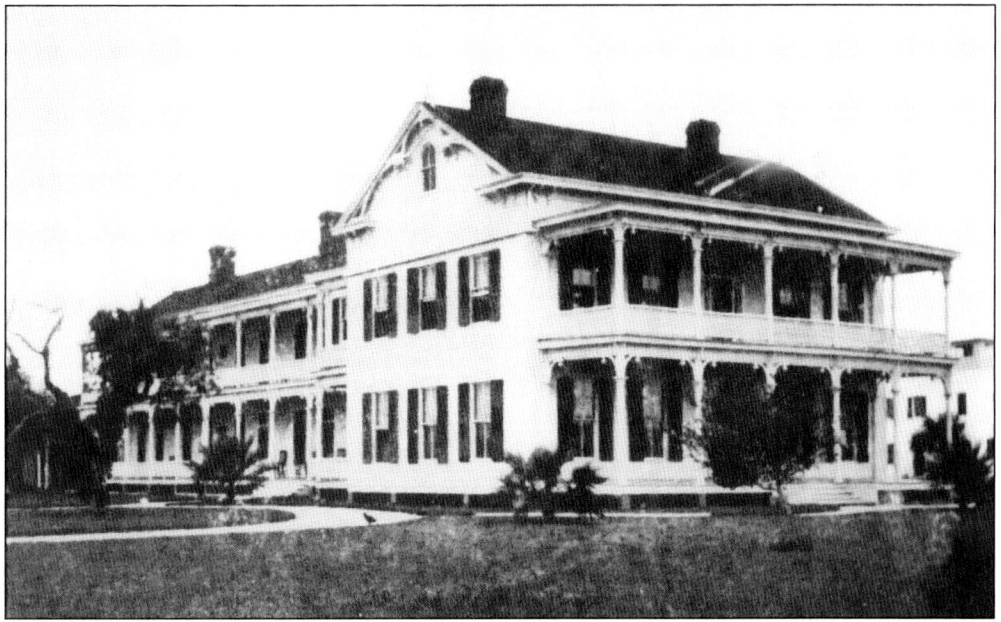

In 1854, the large and rough-mannered King persuaded the petite Henrietta Chamberlain, a preacher's daughter, to marry him. They had five children, and Henrietta King became "La Madama," the beloved *patrona* of the great ranch. The ranch house, seen here, was a rambling frame structure that accommodated countless visitors—the dining table seated 28. Numerous outbuildings included dormitories, shops, corrals, a watchtower, brick cisterns, and a one-room schoolhouse. (Courtesy Historical and Biographical Record of the Cattle Industry of Texas.)

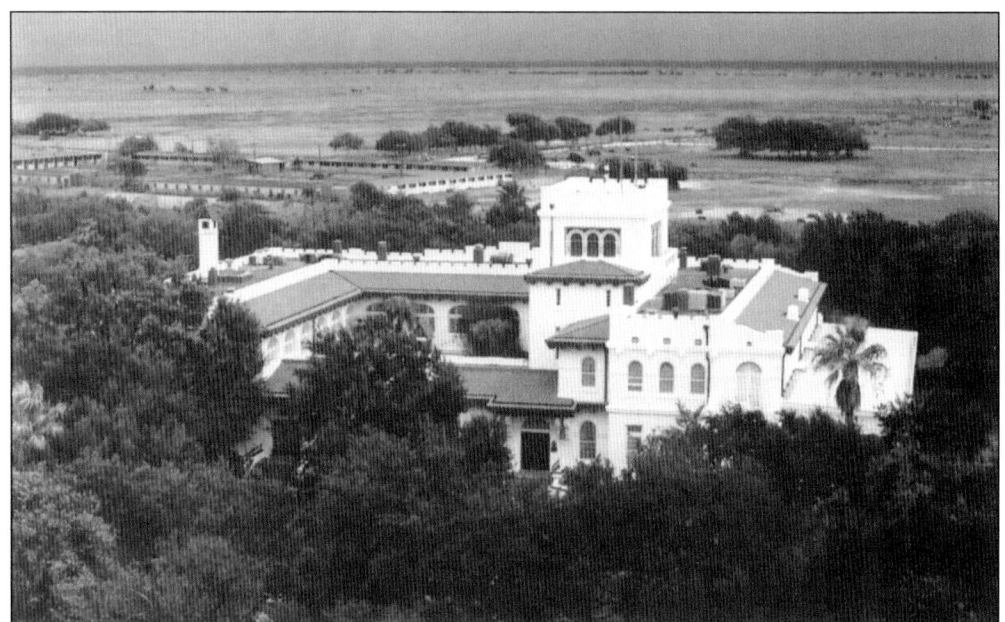

After Richard King died in 1886, Henrietta King continued to direct ranch operations and donned black until her death in 1925. In 1912, the frame ranch house burned and was replaced by this gleaming white 25-room mansion that cost $350,000. (Courtesy King Ranch.)

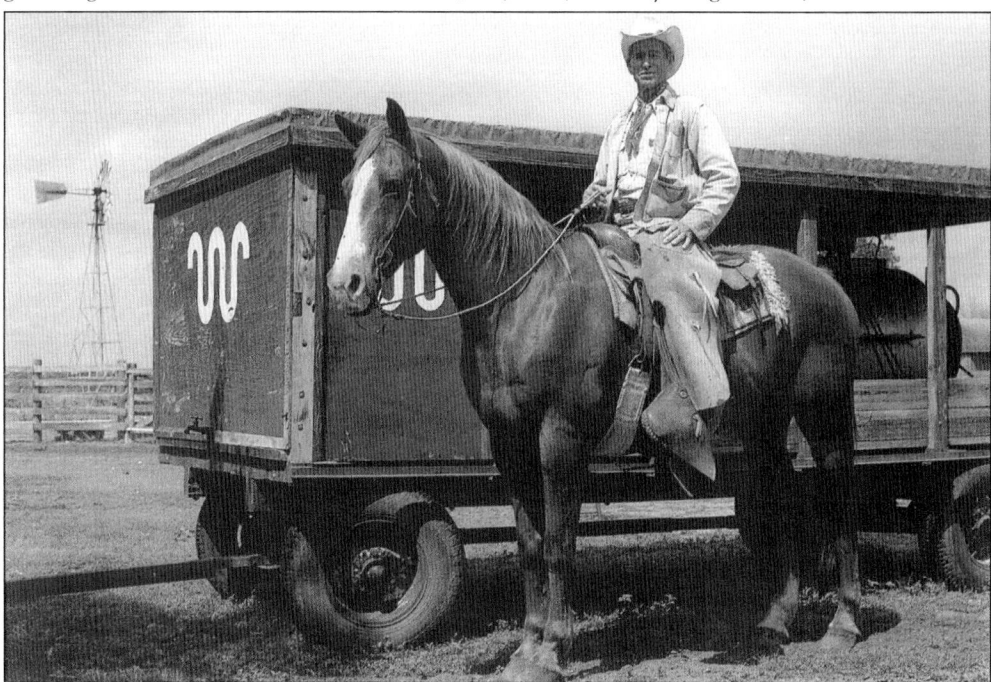

When Richard King was building his original herd, he bought all the cattle in one Mexican village and invited the villagers to migrate north to his ranch. The people formed the nucleus of the *Kiñenos*, who would provide the King Ranch with generation after generation of loyal employees, including Lolo Martinez, seen here, a *Kiñeno* for half a century who provided a direct connection to the Mexican origins of cattle ranching. (Courtesy King Ranch.)

Two

Texas Cattle Trails

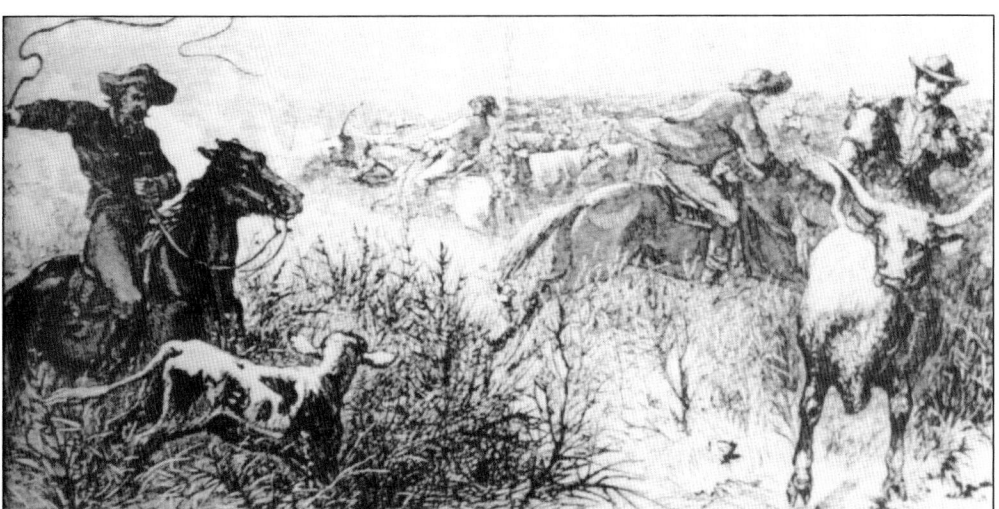

After the Civil War, four million to five million longhorns roamed the rangelands of Texas, most of them in the southern part of the state. Trail crews rode into the brush and rounded up as many wild cattle as they could drive. During the roundup, the unowned animals would be branded. For the first few days, the herd was pushed hard, so they would be tired at night and less likely to stampede. With no purchase price, the cost of wages and provisions divided up to only a few dollars per head, while the cattle might be sold for $30 or $40 apiece. (Courtesy *Harper's Weekly*, May 2, 1874.)

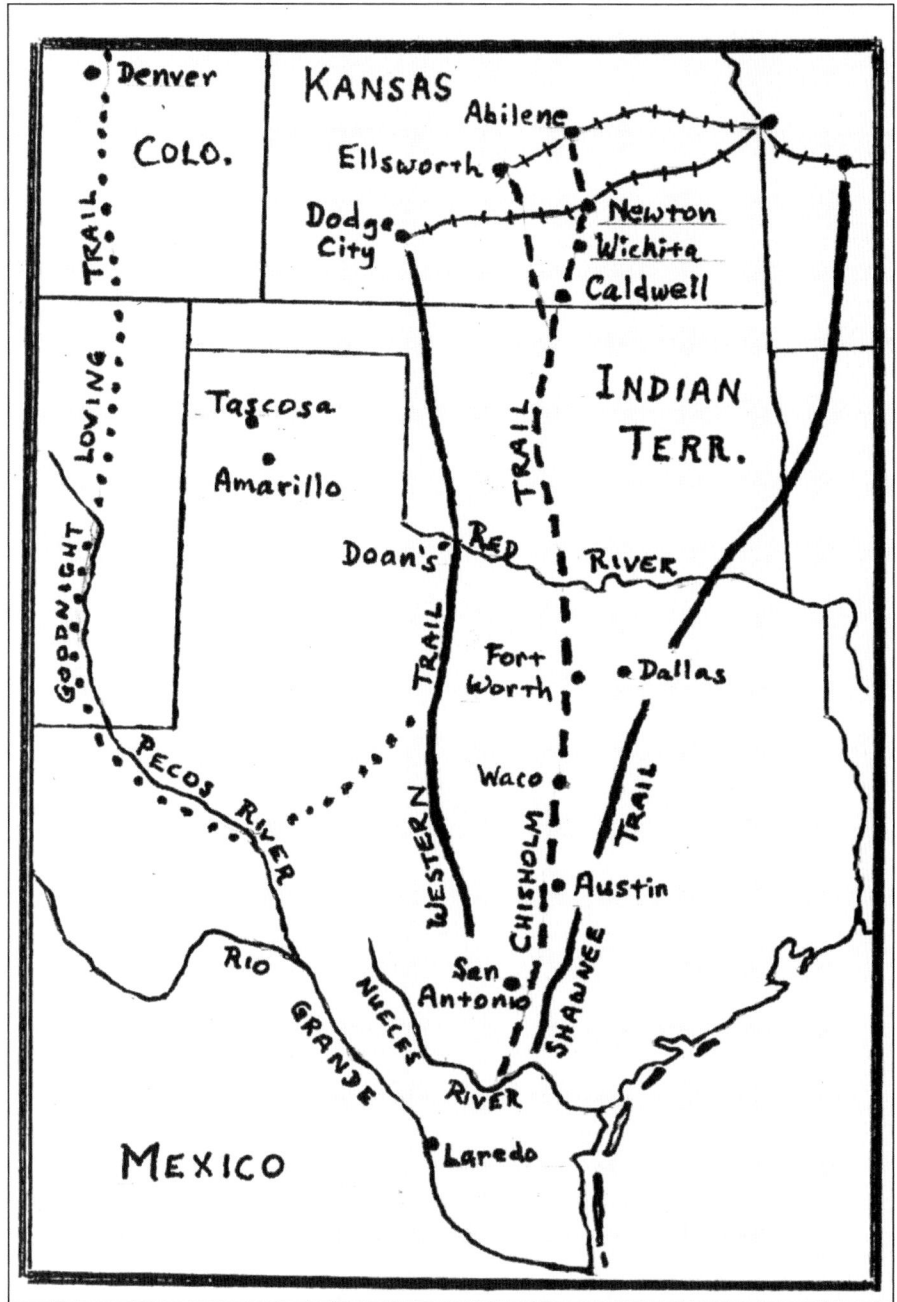

Estimates of Texas cattle driven up the trails after the Civil War run as high as 10 million. The first route, the Shawnee or Sedalia Trail, angles northeast to a rail connection at Sedalia, Missouri. But the route was difficult, through timber and broken country, and Missouri farmers resisted longhorn herds, which spread "Texas fever" through ticks. In 1867, an enterprising cattle buyer, Joseph G. McCoy, established a Kansas railhead at the hamlet of Abilene. The resulting Chisholm Trail became the most famous of all cattle trails, with 600,000 head coming from Texas in 1871 alone. A parallel trail, the Western Trail, pointed to Dodge City. The Goodnight-Loving Trail went across West Texas before heading north along the Pecos River. (Map by the author.)

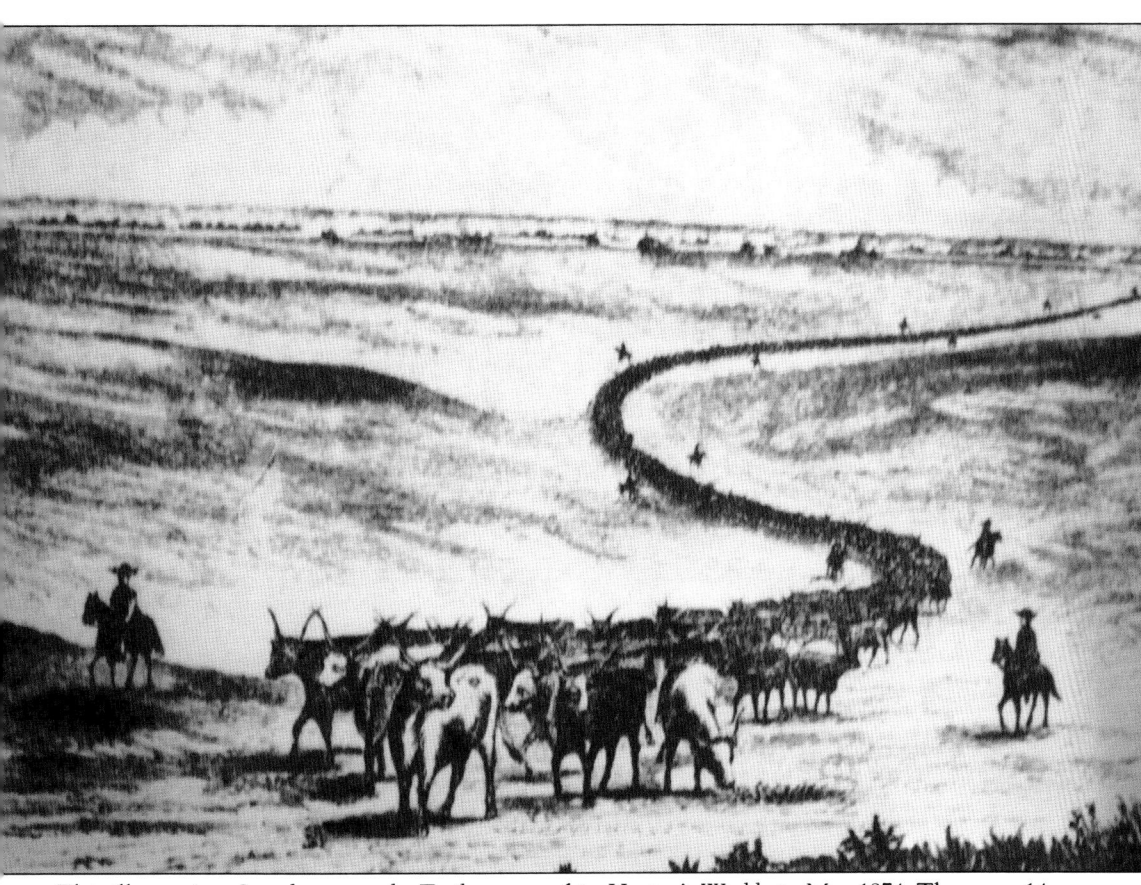

This illustration, *Longhorns on the Trail*, appeared in *Harper's Weekly* in May 1874. There are 14 riders in the illustration, an unusually large crew. The two men in the foreground are riding point. Spaced behind them are six drovers riding on each flank. Three or four men would be out of sight to the right, eating dust as they rode drag, pushing the slowest and laziest cattle. Drovers rotated positions from day to day, because no one wanted to ride drag all the way to Kansas. The trail boss led the way (out of sight to the left), the cook drove the chuck wagon ahead of the herd, and the horse wrangler herded the remuda in the distance. (Courtesy *Harper's Weekly*, May 2, 1874.)

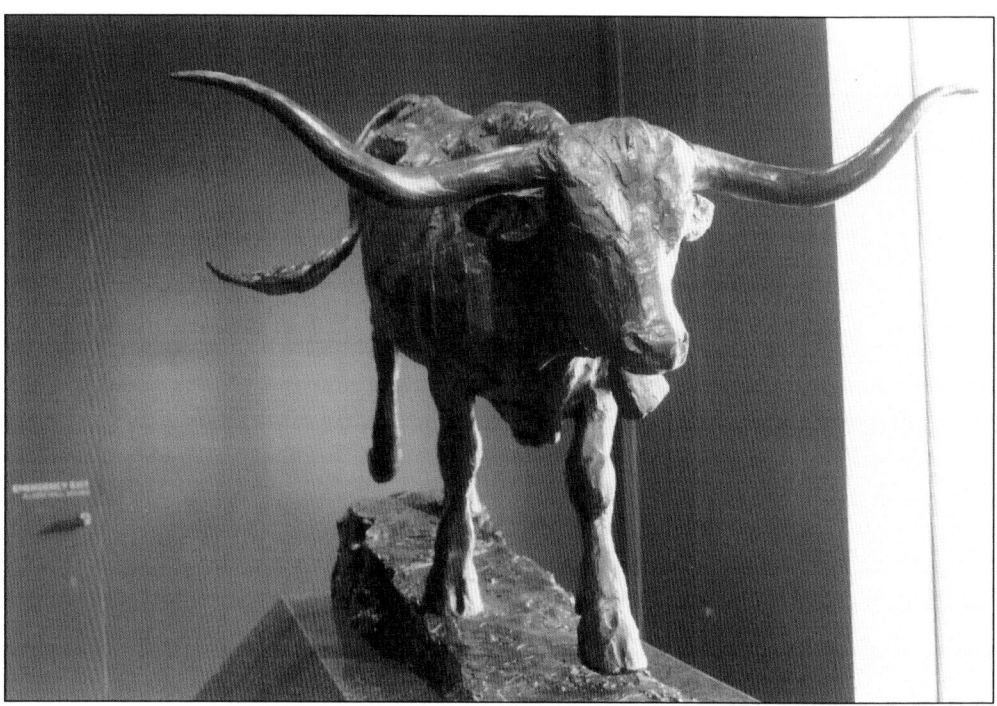

Dominant steers forged to the head of trail drives and led the herds. The most famous lead longhorn was Old Blue, a 1,400-pound JAs steer who was used again and again by Charles Goodnight as a trail herd leader. When a trail drive commenced, a brass bell was attached to Old Blue's neck. The herd animals became so accustomed to following the bell and the tireless strides of Old Blue that at night the bell was muffled, since the sound of the clapper would bring the herd to its feet, ready to travel. Old Blue was shipped back to the ranch after each drive, and eventually he was turned out to pasture for an honorable retirement. There is an imposing statue of Old Blue at the Texas and Southwestern Cattle Raisers Association Museum in Fort Worth (above), and the horns are displayed at the Panhandle Plains Museum in Canyon (below). (Photographs by author.)

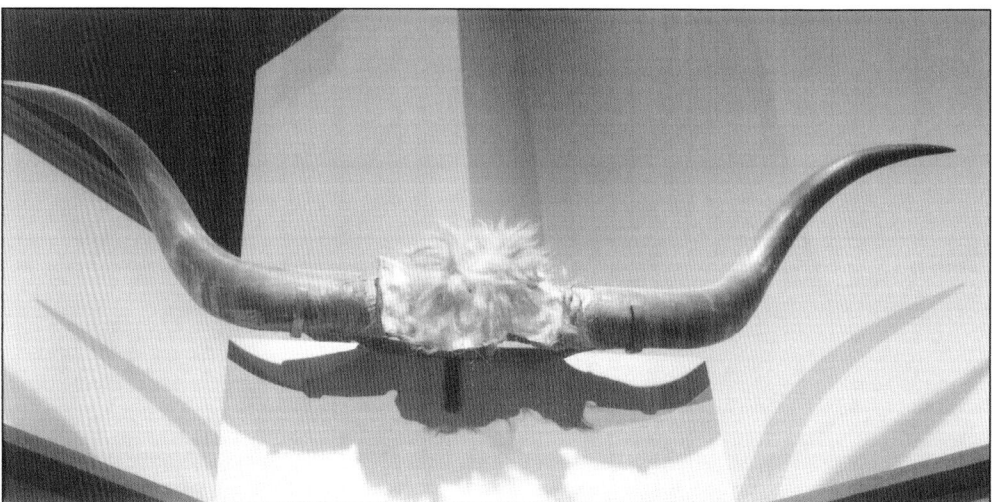

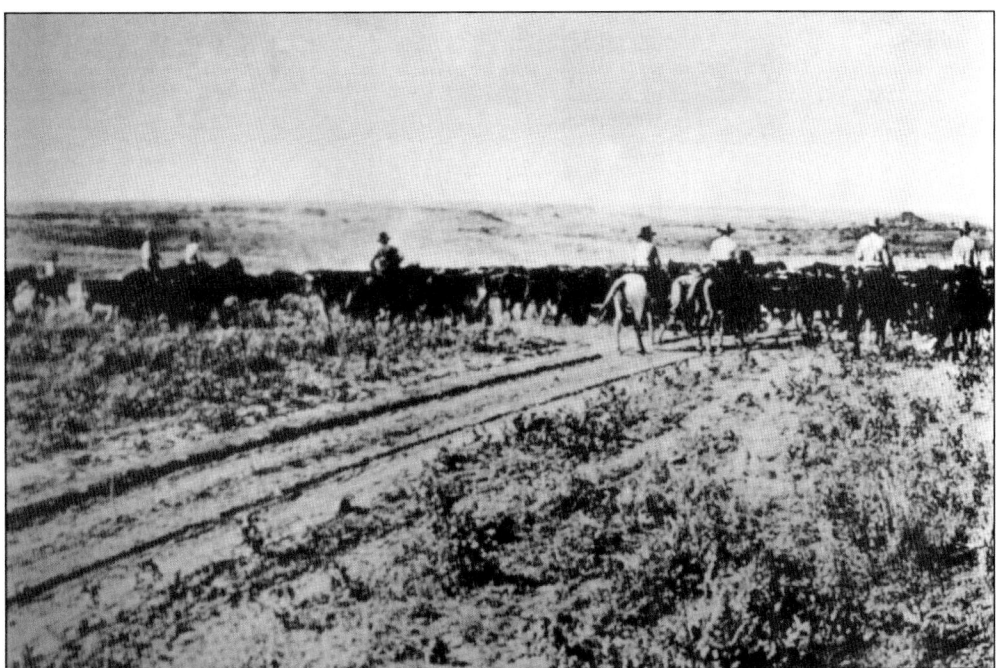
Weaker or lazy cattle that brought up the rear were called drags, and the drovers who pushed them, seen here, were drag hands. (Courtesy Historical and Biographical Record of the Cattle Industry of Texas.)

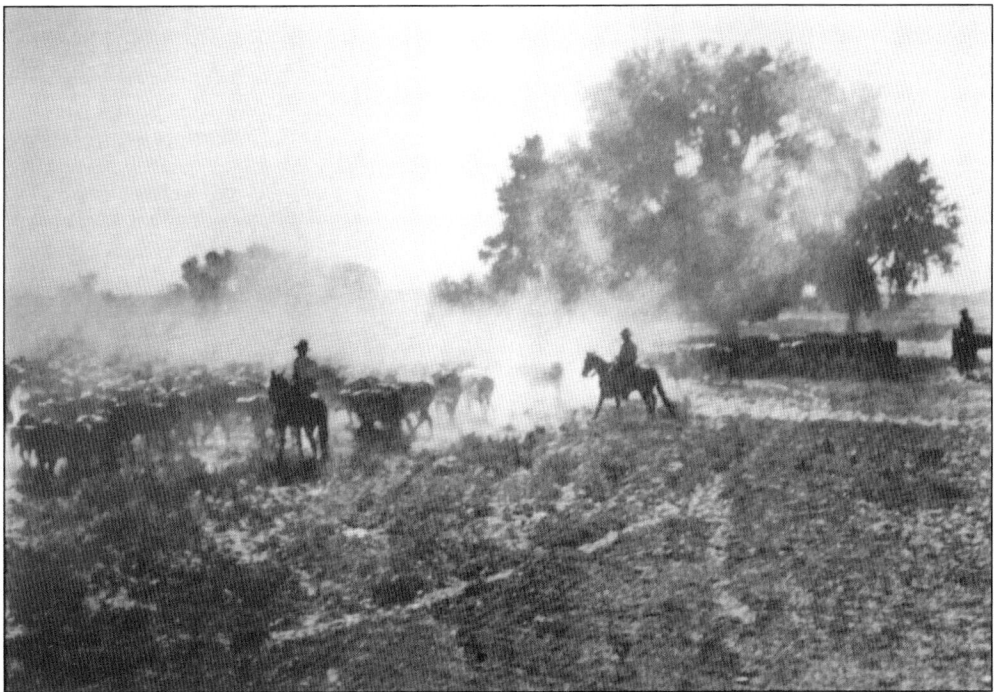
Charles Goodnight said that he "always selected three steady men for the rear." Riding drag was dusty work and, added Goodnight, "the heat from many moving cattle was terrific." (Courtesy Historical and Biographical Record of the Cattle Industry of Texas.)

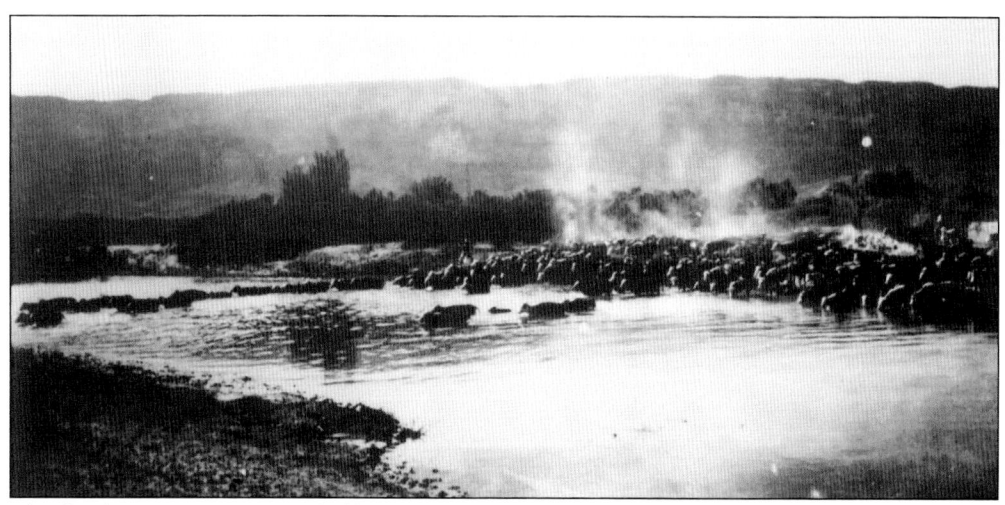
This herd crosses a river at a shallow ford. (Courtesy Historical and Biographical Record of the Cattle Industry of Texas.)

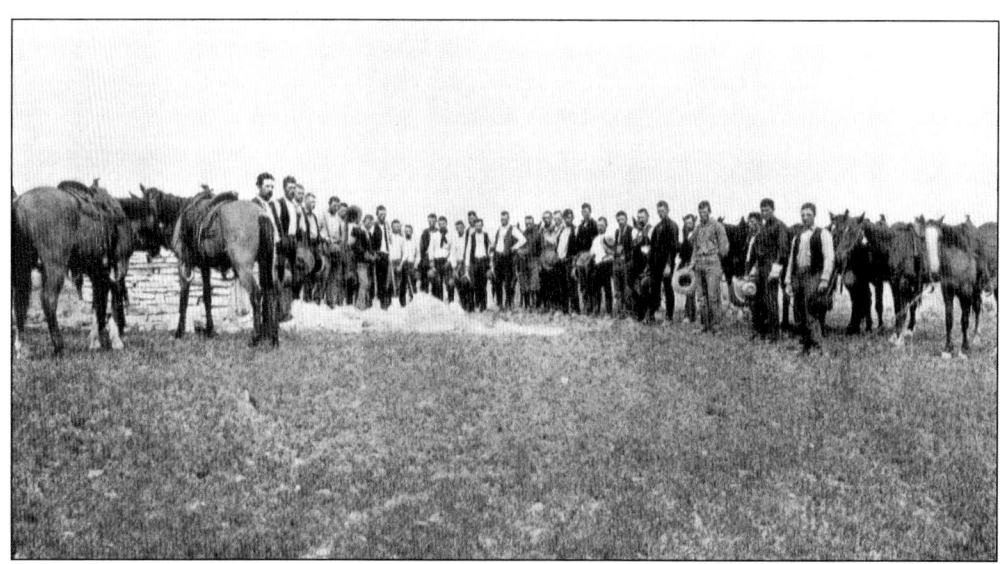
Cowboys sometimes drowned during river crossings or were trampled to death during stampedes. Here, 30 men of the range have gathered to bury an unlucky comrade "on the lone prairie." (Courtesy Historical and Biographical Record of the Cattle Industry of Texas.)

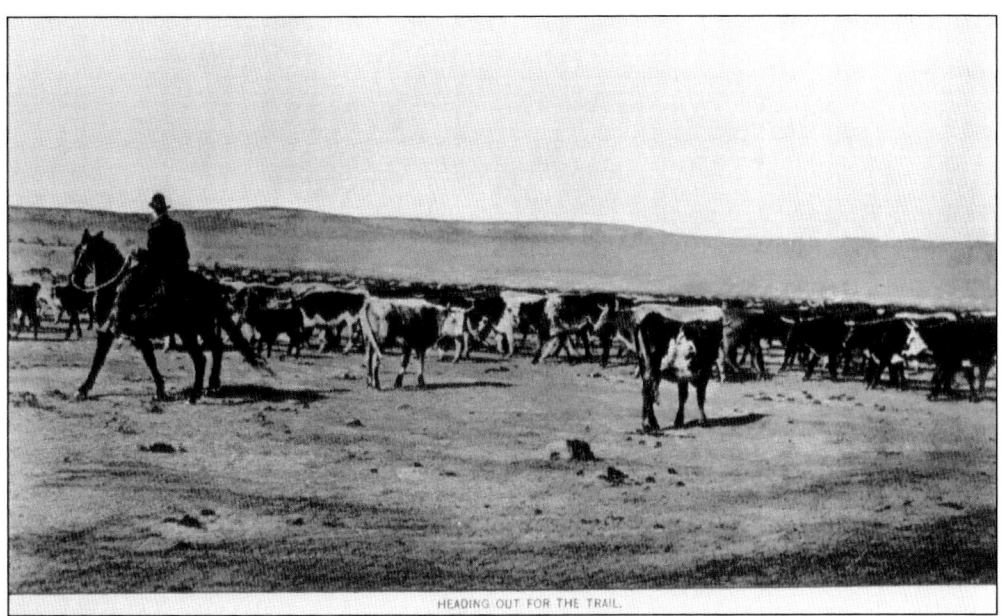

This photograph, with the explanatory title *Heading Out For the Trail*, shows the start of a trail drive. (Courtesy Historical and Biographical Record of the Cattle Industry of Texas.)

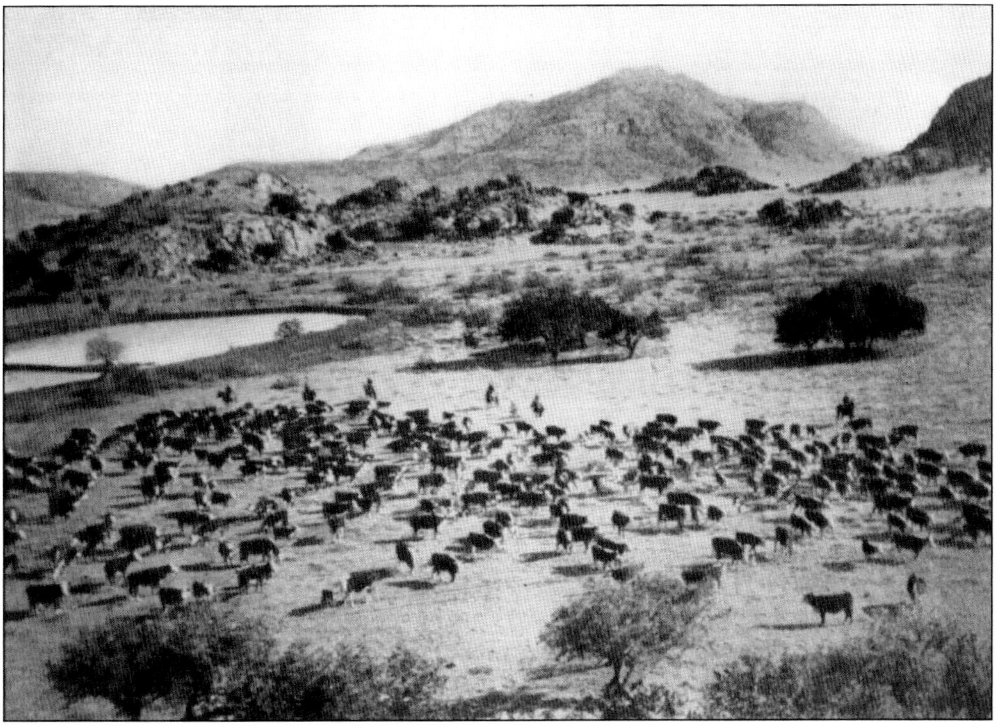

Following the noon grazing, according to Charles Goodnight, a trail herd "would not eat any more until they got to water, which we always tried to reach before sundown." He continued, "This gave us ample time to have the cattle filled and everything arranged for a pleasant night. After they had grazed they bedded down for the night. The herd was put in a circle, the cattle being a comfortable distance apart." (Courtesy Library of Congress.)

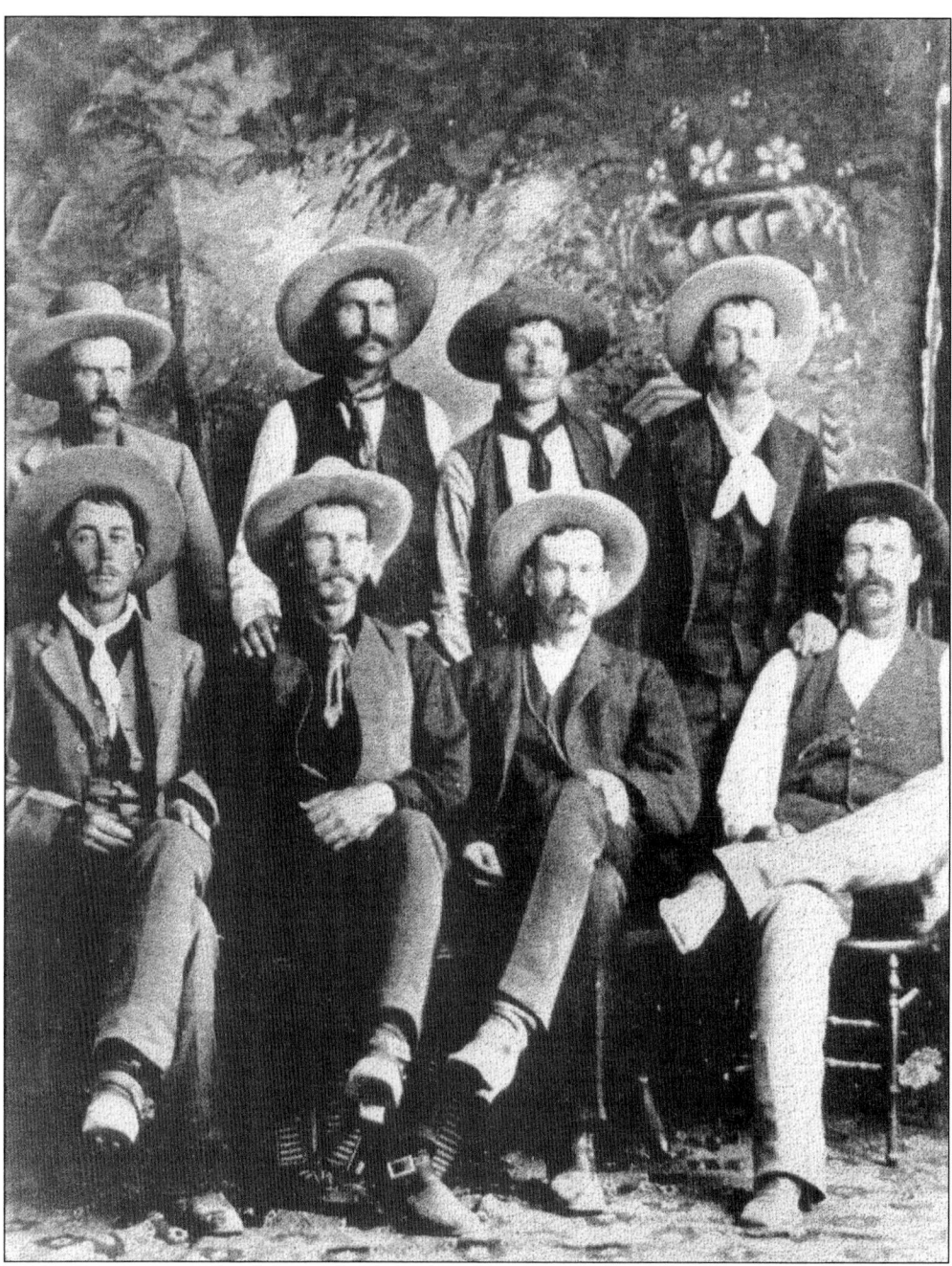

Trail boss Pink Higgins (seated, far right) and a crew of drovers from Lampasas County pose for a photograph after driving a herd to Kansas. Higgins went on his first cattle drive in 1868 when he was 17, helping to shove a herd to Cheyenne, Wyoming, "The Holy City of the Cow." Higgins and his father, John, owned a ranch in Lampasas County, and within a few years, he was a trail boss. One of his most experienced drovers, Alonzo Mitchell (standing, far right), observed, "Chain lightning caused more stampedes than anything else, and next came lobo wolves—the smell of them." Another veteran drover was Jess Standard (seated, second from left), the author's great-grandfather. (Author's collection.)

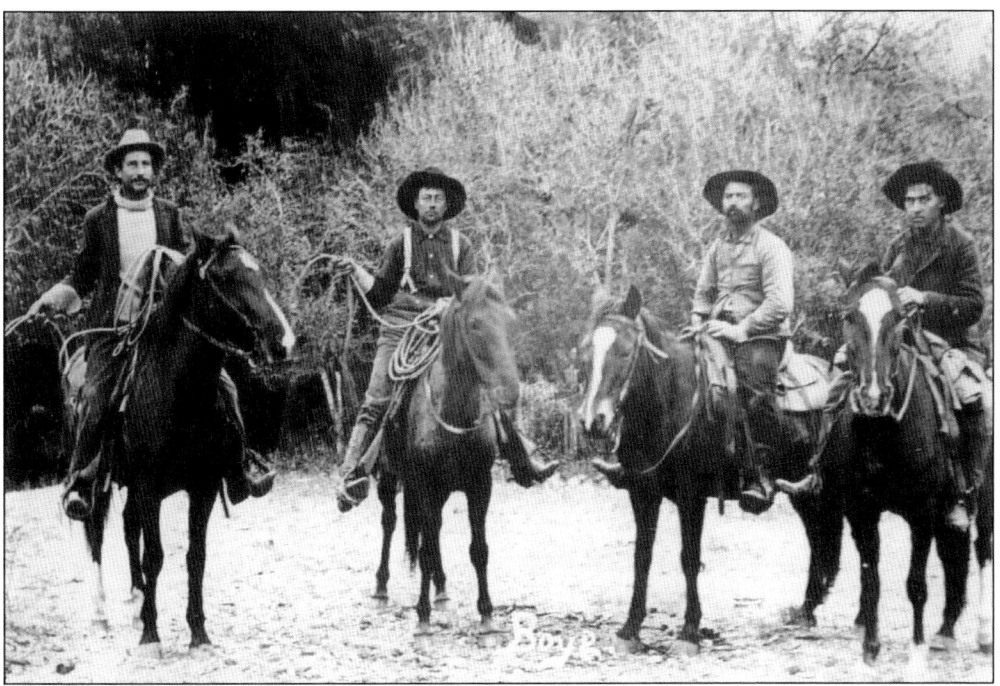

Two of these men show off their lassoes for the photographer. Charles Goodnight once reminisced, "I wish I could find words to describe the companionship and loyalty of some of the men towards each other. It is beyond imagination." (Courtesy Scurry County Historical Museum, Snyder.)

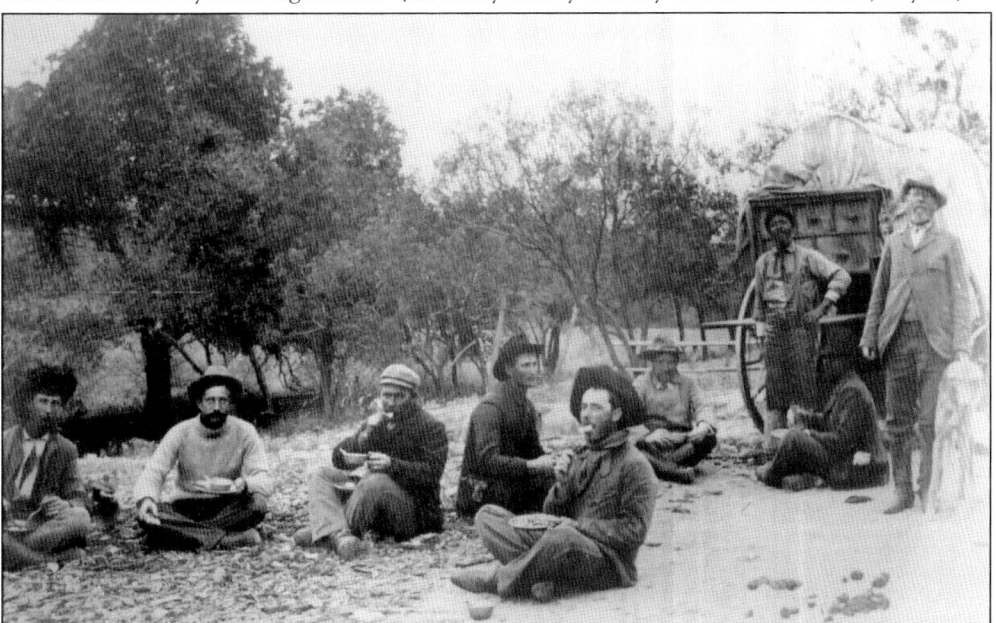

The cook usually prepared lunch while the crew ate breakfast. The chuck wagon then went ahead of the herd and set up at the grazing ground that the trail boss selected for the midday stop, which usually began about 11:00 a.m. After dinner, the chuck wagon went ahead to the evening water and bed ground, where the cook prepared the supper. (Courtesy Scurry County Historical Museum, Snyder.)

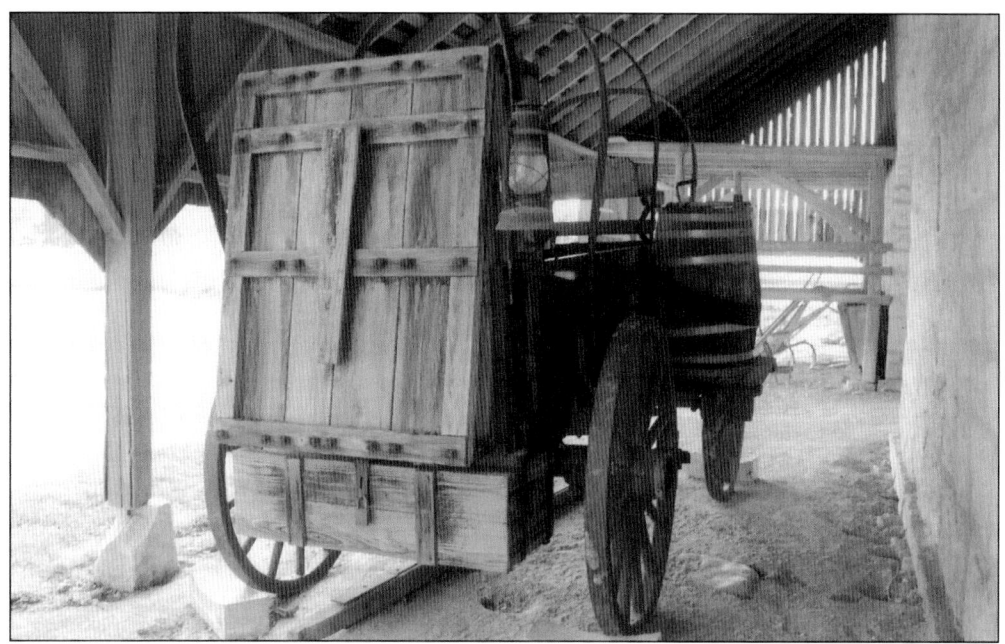

This chuck wagon is now on display at the old Johnson ranch headquarters at Johnson City. The lid to the chuck wagon is up, ready for travel. Note the lantern and the water barrel. The wagon bed hauled sacks of flour and beans, the bedrolls of the crew, slickers, rope, an extra wagon wheel, kerosene for the lantern, guns and ammunition, an ax, a shovel, and hobbles. (Photograph by the author.)

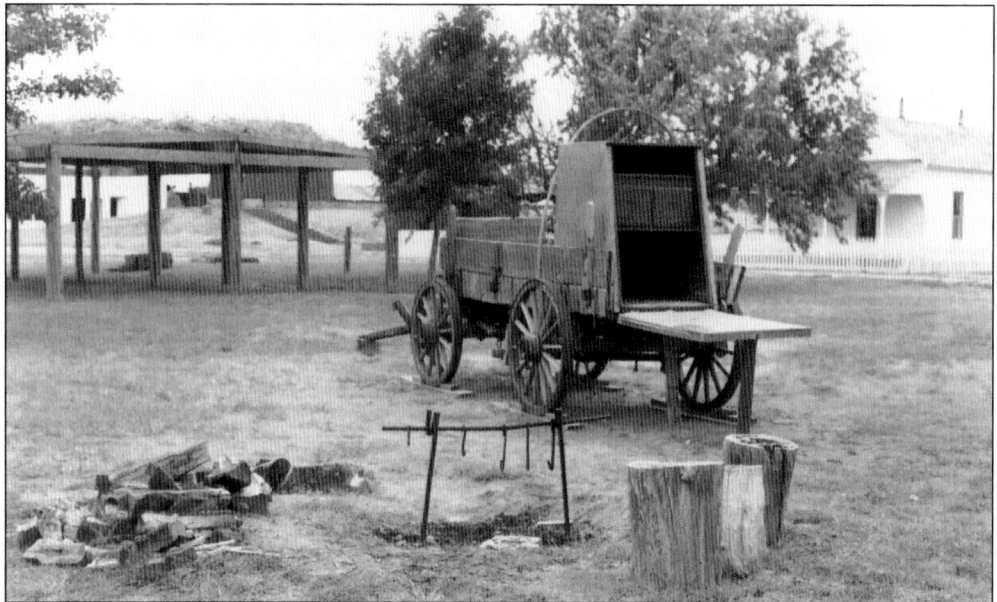

The lid is down, creating a worktable on this chuck wagon at the Ranch Heritage Center in Lubbock. The drawers and shelves of the chuck box would carry plates, cups, cutlery, tobacco, matches, sugar, dried fruit, coffee, a big coffee pot, salt, lard, baking soda, molasses, sourdough, vinegar, bandages, castor oil, and whiskey. The boot, or lower part of the chuck box, contained skillets and Dutch ovens. (Photograph by the author.)

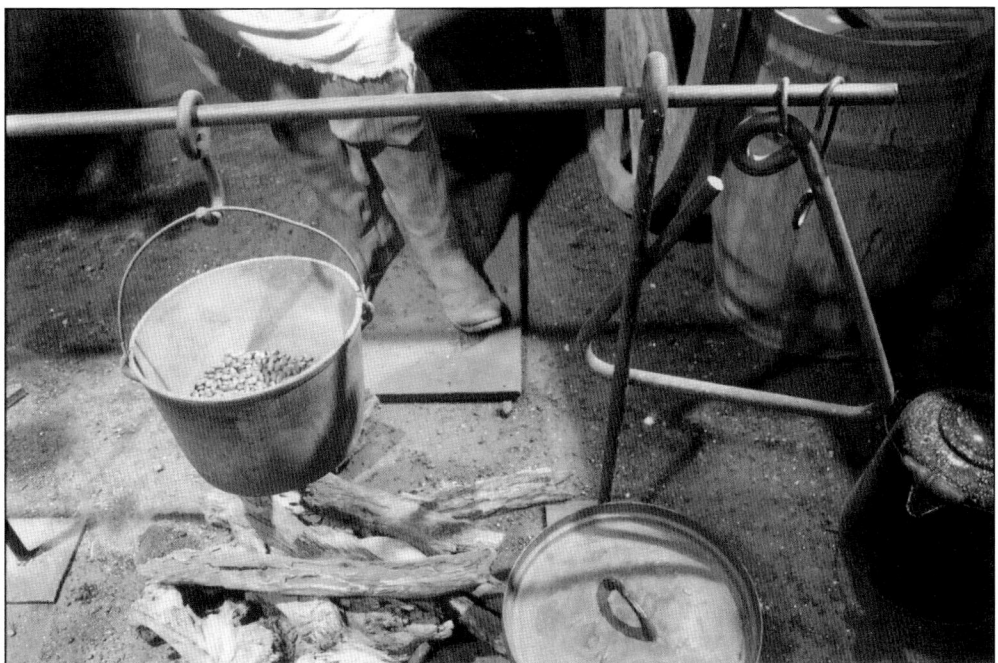

Almost every chuck wagon meal on a trail drive featured "Pecos strawberries" (beans), "overland trout" (bacon), and sourdough biscuits. This photograph was taken at the Scurry County Museum in Snyder. (Photograph by the author.)

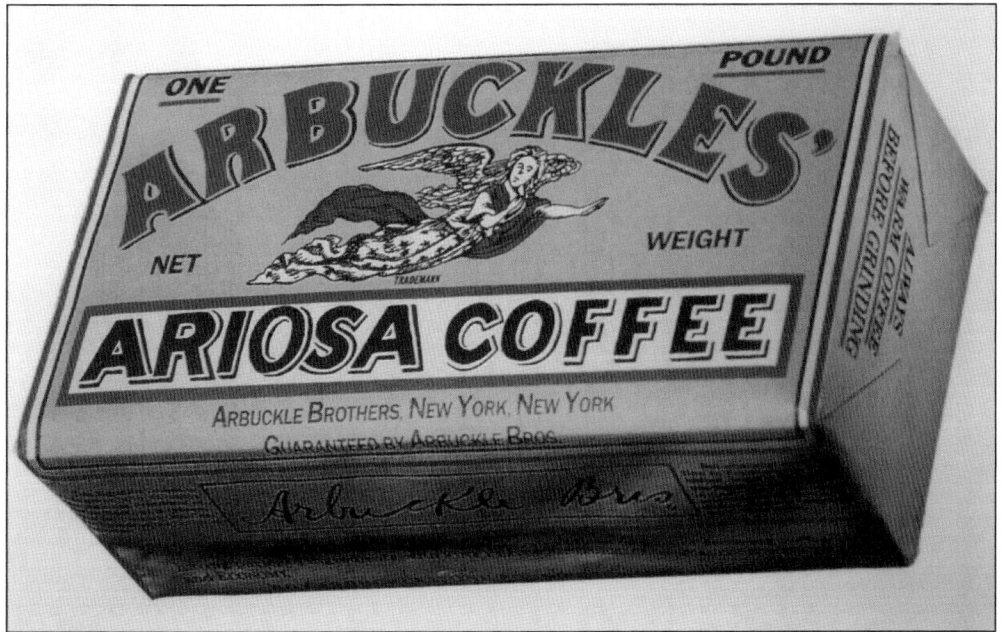

Arbuckle's was so popular on the cattle frontier that many cowmen did not know there was another brand of coffee. Chuck wagon cooks liked to brag about how to make strong coffee. Ramon Adams notes in *Western Words*: "Take two pounds of Arbuckle's, put in enough water to wet it down, boil for two hours, then throw in a horse shoe. If the horse shoe sinks, she ain't ready." (Author's collection.)

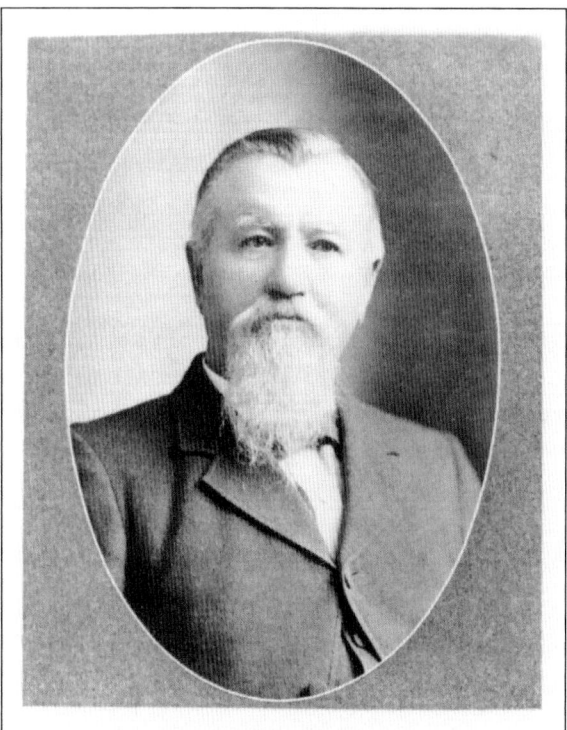

JOHN T. LYTLE

A native of Pennsylvania, John Lytle came to Texas at the age of 15 in 1860 to cowboy on his uncle's ranch west of San Antonio. Following three years in the Confederate army, Lytle began ranching and trail driving on his own. In 1871, he blazed the Western Trail, from Fort Griffin to Dodge City, and over the next 16 years, he sent 450,000 cattle up the trail. (Author's collection.)

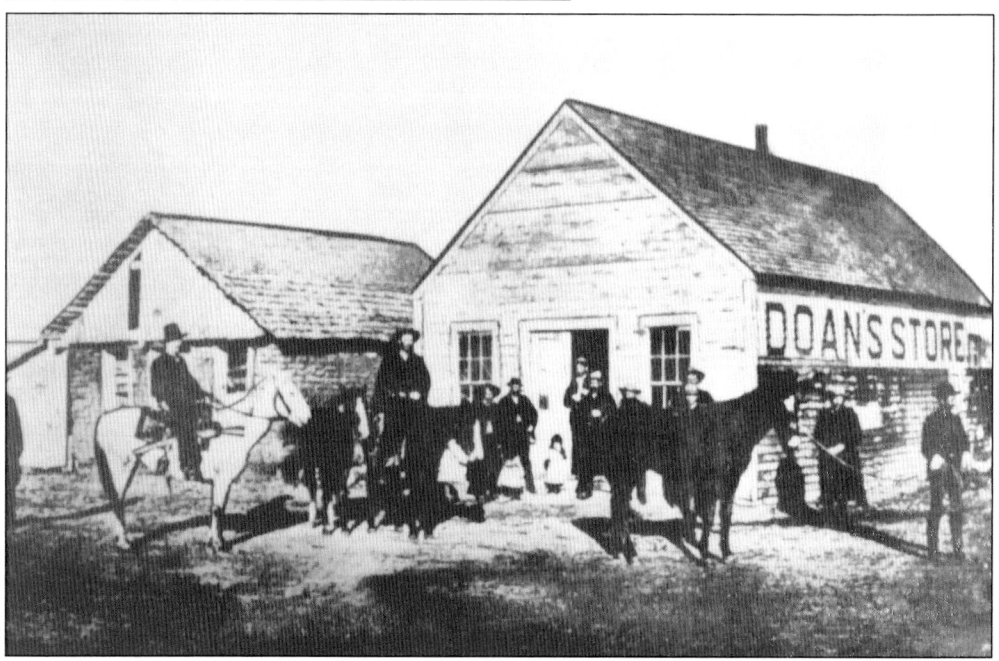

From Fort Griffin, herds were trailed north to Doan's Crossing at the Red River. The village had more than a dozen buildings, most notably the store of Judge Corwin F. Doan. The store sold whiskey, guns, blankets, and saddles, and Doan and his wife, Lide, entertained leading cattlemen at their nearby adobe house. (Courtesy Historical and Biographical Record of the Cattle Industry of Texas.)

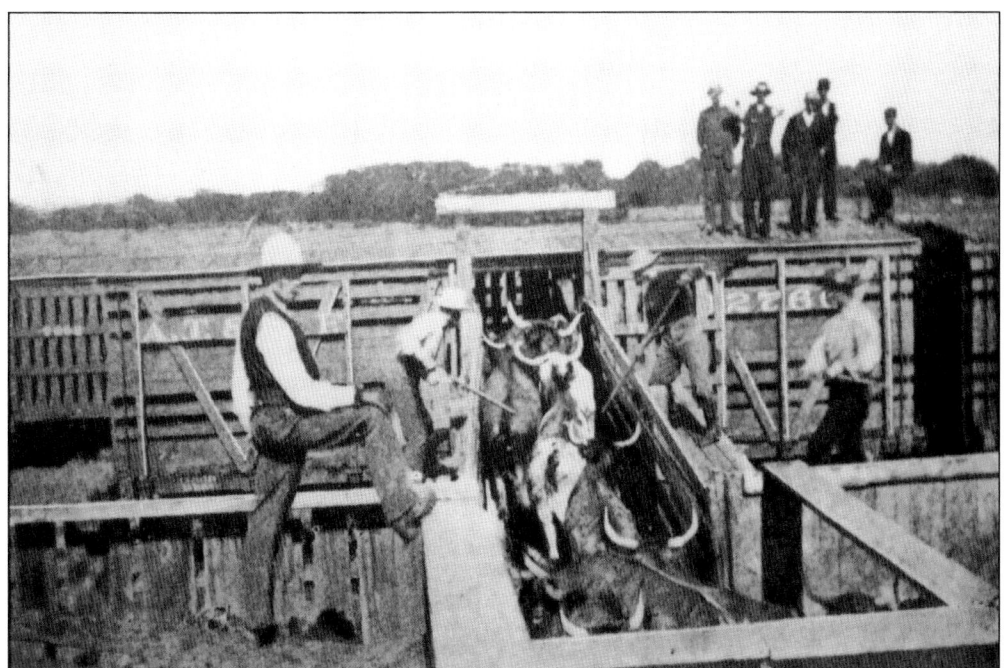

Texas cattle were loaded onto railroad cars at the shipping pens just south of Caldwell. The men prodding the cattle along demonstrate the origin of the term "cowpoke." (Courtesy Caldwell History Association.)

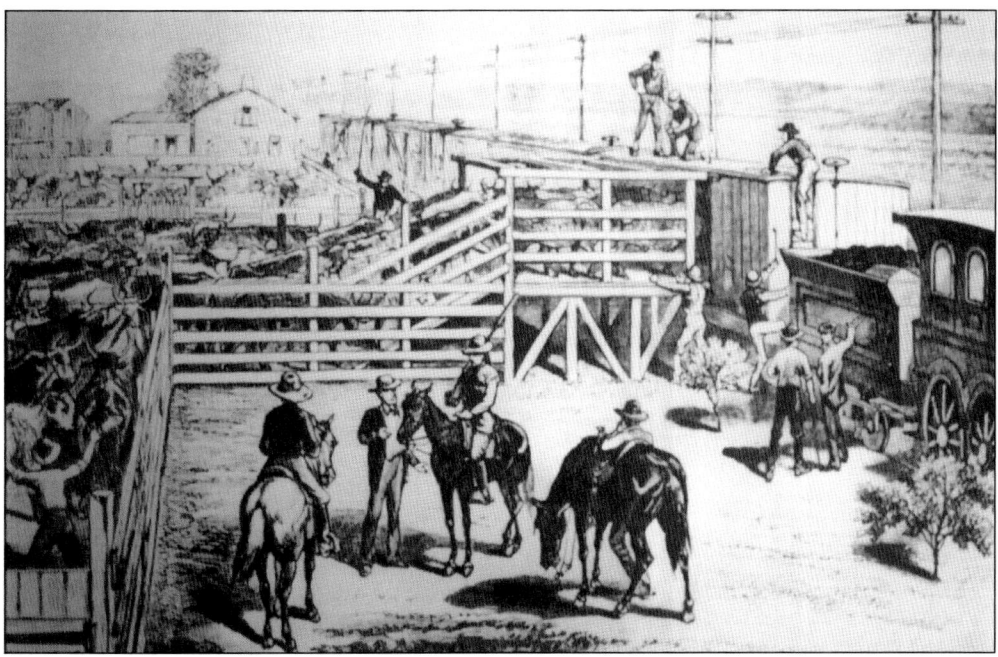

On September 5, 1867, the first cattle were shipped out of Abilene to Chicago. The new stockyards had just been erected by enterprising cattle buyer Joseph G. McCoy, who also built barns, livery stables, and the Drover's Cottage, a three-story frame hotel. (Courtesy Historical and Biographical Record of the Cattle Industry of Texas.)

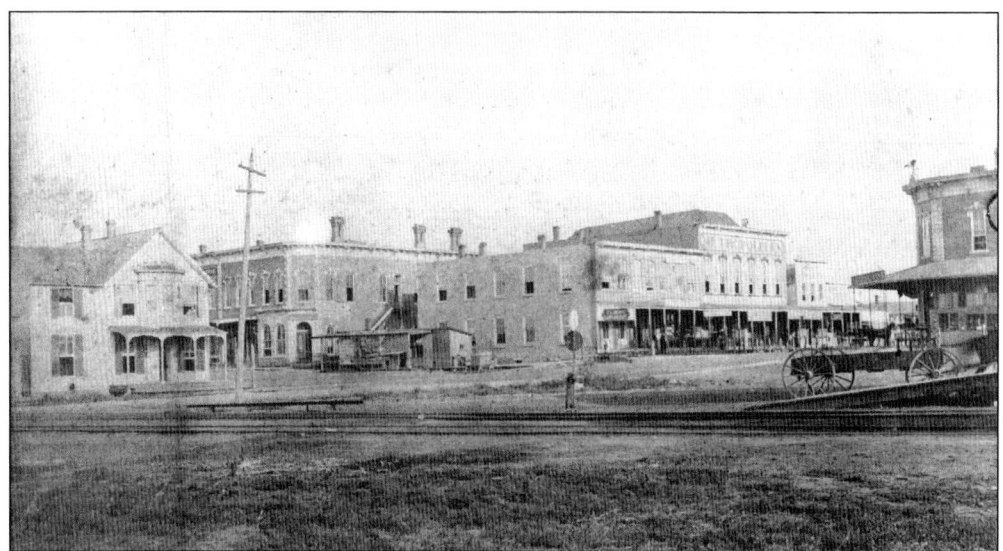

Abilene, Kansas, was the first Wild West town where Texas cowboys engaged in mass drinking and brawling sprees, thereby establishing an indelible image. At the end of a long drive, cowboys swaggered along Texas Street into the Alamo Saloon, the Lone Star Saloon, and the Bull's Head Saloon. Dusty drovers bathed, bought new clothes, drank, gambled, visited prostitutes, and often blew four months' wages in a few raucous days. (Courtesy Kansas State Historical Society, Topeka.)

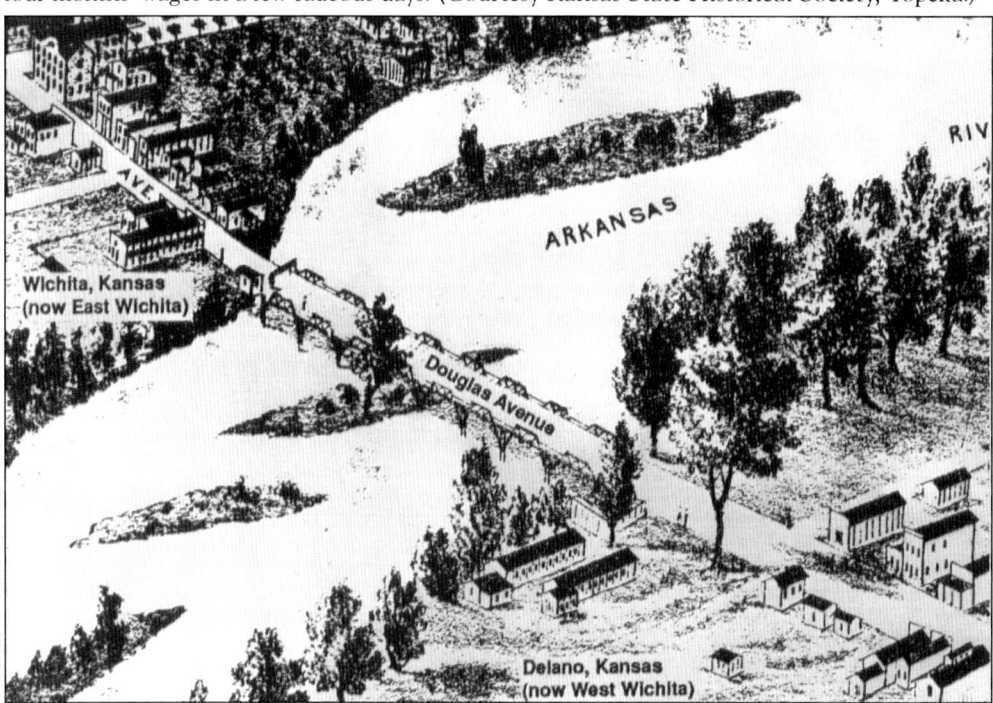

By 1872, a railroad line reached Wichita, which then became the terminus of the Chisholm Trail. The commercial center of town was at the intersection of Douglas and Main Streets. West on Douglas Street and across the Arkansas River Bridge was the red light district. The two most notorious dives were the dance halls of "Rowdy" Joe Lowe and "Red Beard," who was killed in a shoot-out with Lowe in 1873. (Author's collection.)

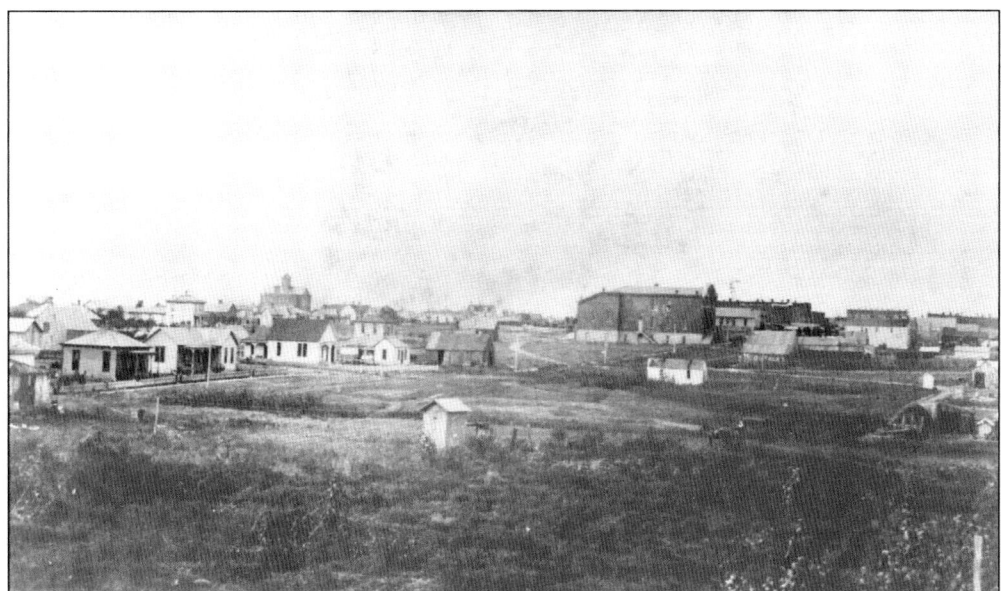

An unincorporated village just above the southern border of Kansas, Caldwell was the first community encountered by drovers emerging from Indian Territory. For a decade, the lawless village was a wild trail town. When the railroad was extended south from Wichita in 1880, Caldwell became the last railhead on the famous trail, boasting two three-story brick hotels and a brick opera house (right of center). (Courtesy Kansas State Historical Society, Topeka.)

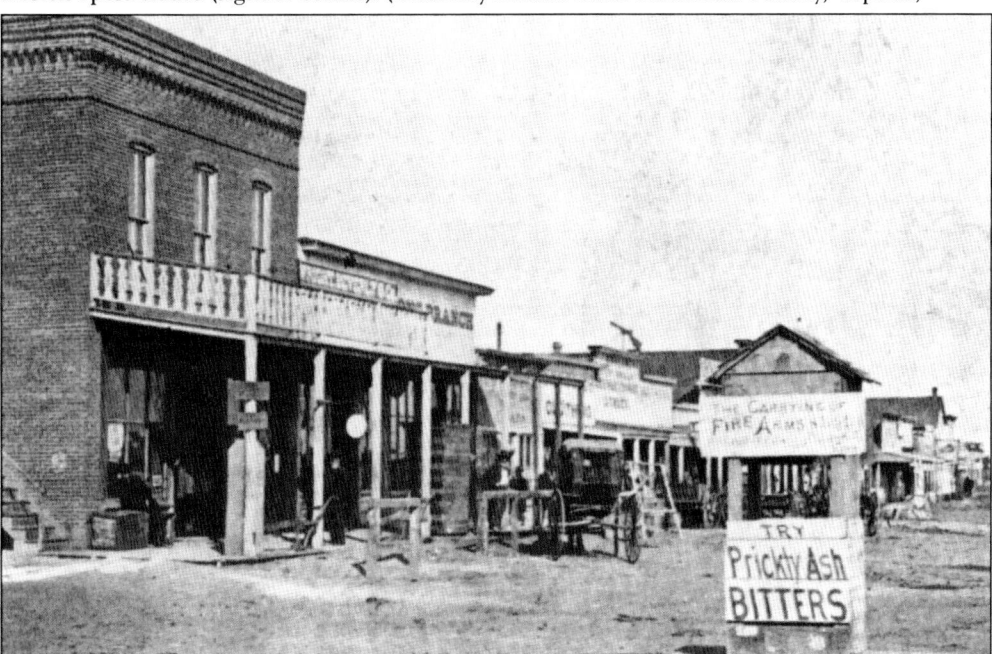

Dodge City was notorious as the "Bibulous Babylon of the Plains." The two-story brick building seen here on Dodge City's Front Street was Wright, Beverly & Co., next to the famous Long Branch Saloon. Doc Holliday and Luke Short gambled in Dodge, while Bat Masterson, Bill Tilghman, and Wyatt and Morgan Earp wore badges. City marshal Ed Masterson died in a wild gunfight. (Courtesy Kansas State Historical Society, Topeka.)

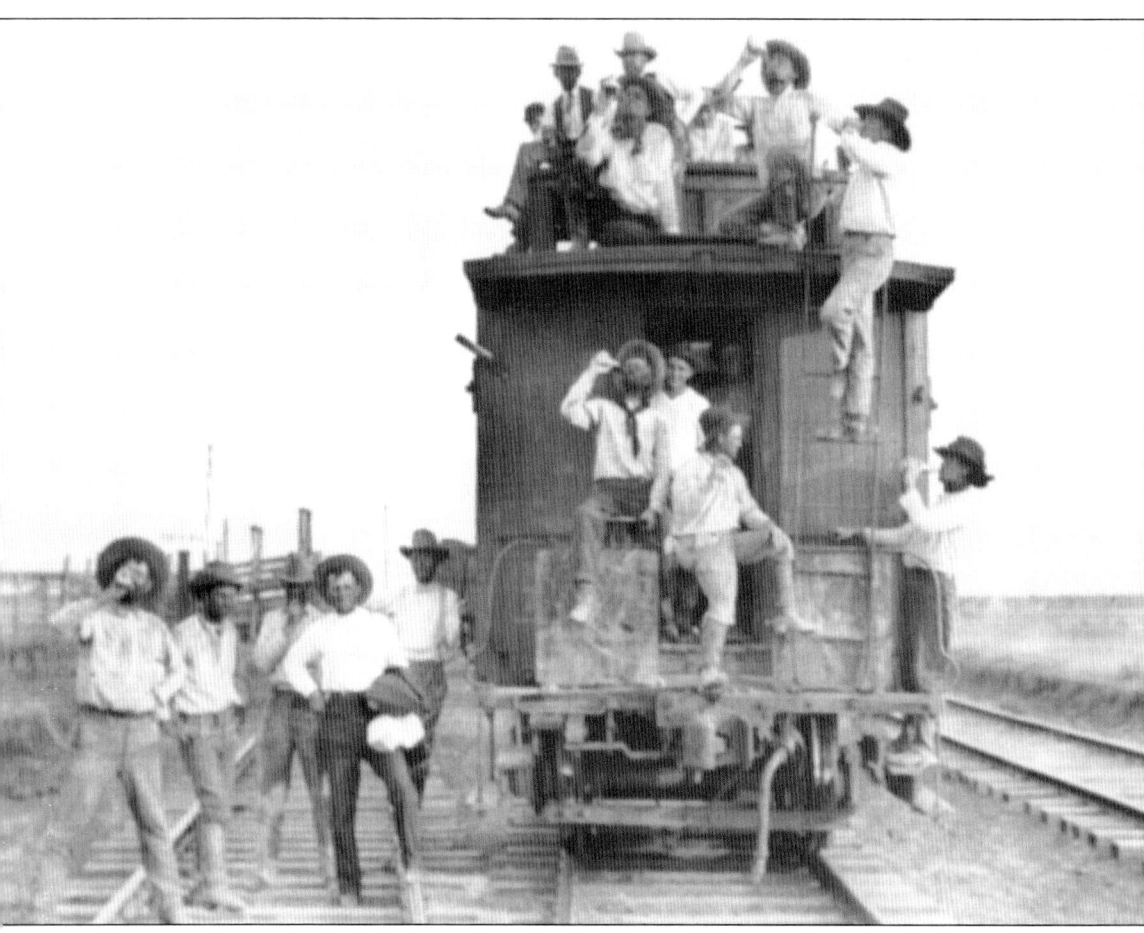

Large ranches maintained their own horse herds. On a trail drive, each cowboy had several horses, but at the end of the drive, this remuda was driven back to the home ranch. These Spur Ranch cowboys are being taken back by train, often along a circuitous route. But the Spurs cowboys have not ended their end-of-drive celebration, and the train ride promises a jolly time. (Courtesy Panola College Library, Carthage.)

Three
Charles Goodnight

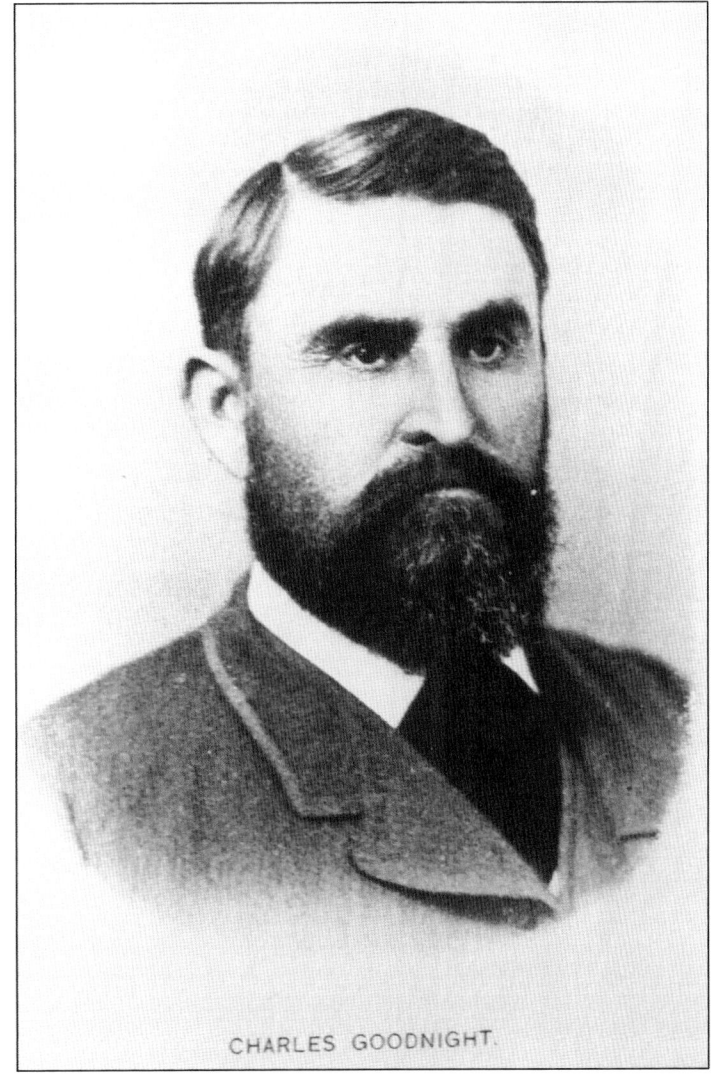

Charles Goodnight was a legendary cattleman of the Texas frontier. Renowned as a trailblazer, Goodnight opened five cattle trails after the Civil War, the most famous of which was the Goodnight-Loving Trail. He had an uncanny sense of direction and terrain, and his masterful organization and careful precautions earned him safe passage where others encountered a variety of disasters. To facilitate his trail drives, Goodnight invented the chuck wagon, which quickly became an iconic feature of the range cattle industry. (Author's collection.)

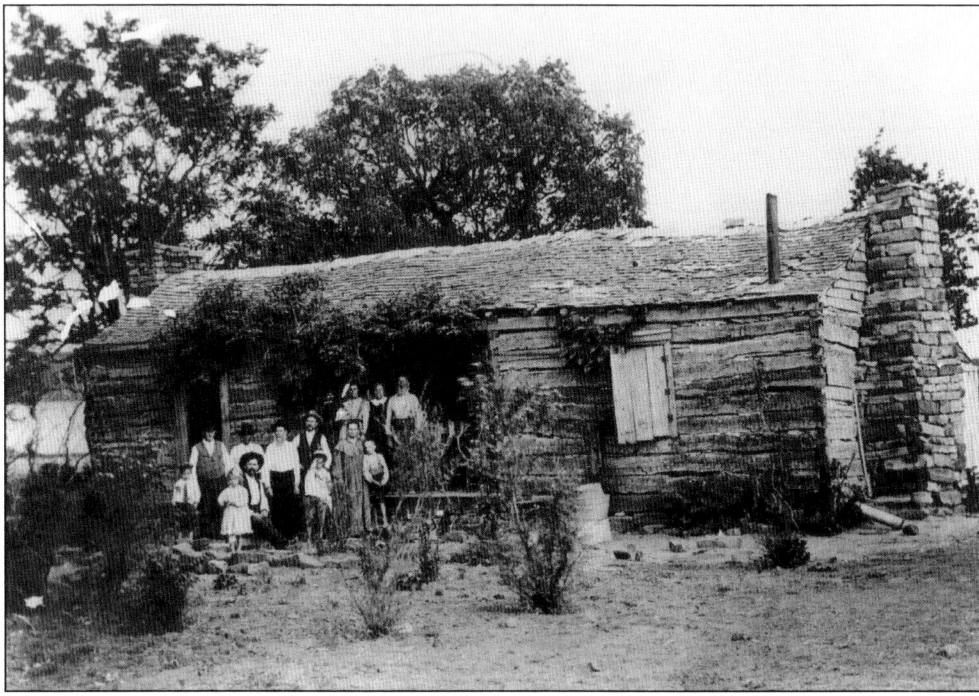

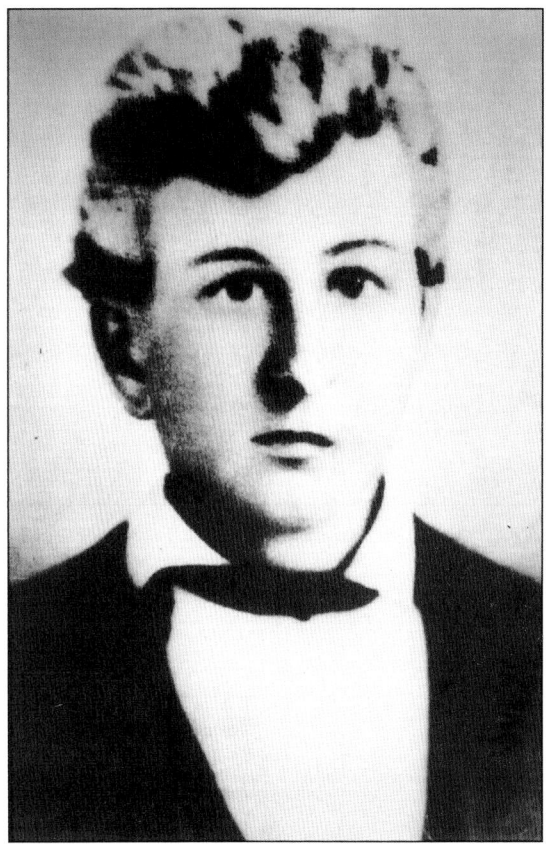

Charlie Goodnight was born in 1836 in Illinois to Charles and Charlotte Goodnight. In 1845, the family moved to the Republic of Texas, and nine-year-old Charlie rode 800 miles bareback on his first frontier adventure. By the time he was 20, Goodnight was running cattle on the frontier of northwest Texas. For the next decade, he centered his ranching activities in Palo Pinto, Parker, and Young Counties. He built the substantial log cabin above for his widowed mother. (Courtesy Panhandle Plains Historical Museum, Canyon, Texas.)

Born in Kentucky in 1812, Oliver Loving (left) migrated to the Republic of Texas in 1843 with his wife, Sarah, and their growing family. By 1855, Loving was ranching in what would become Palo Pinto County. In 1857, he sent the first Texas cattle herd to Chicago, earning a profit of $36 a head. Loving began to trail cattle to the Colorado mining district and provided beef to the Confederate army during the Civil War. (Author's collection.)

In 1866, partners Charles Goodnight and Oliver Loving blazed the Goodnight-Loving Trail from northwest Texas west to the Pecos River and then north into New Mexico and beyond. The next year, Loving was wounded by a war party along the trail. His shattered arm turned gangrenous, and on his deathbed at Fort Sumner, New Mexico, he asked to be laid to rest in Texas. Goodnight brought his remains to Weatherford. (Photograph by the author.)

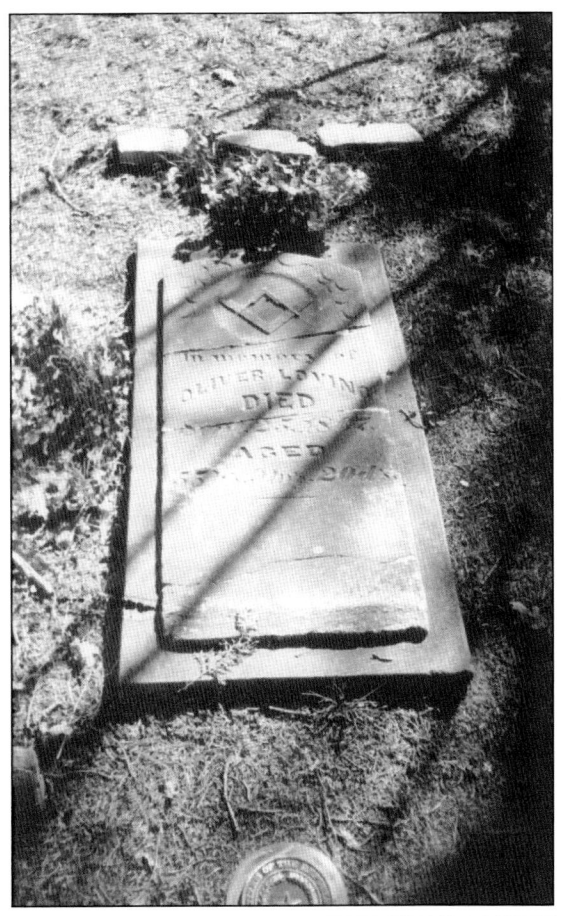

For decades, Palo Duro Canyon (below) offered Comanches a winter refuge from both weather and enemies. Half a mile or so deep and 100 miles long, the canyon was sheltered from weather, watered by the Prairie Dog Town Fork of the Red River, and populated by buffalo, the Comanche staff of life. But when the Army finally discovered the canyon in 1874, Comanches were driven onto the reservation. And within a year, Charles Goodnight drove 1,600 head of cattle into Palo Duro Canyon. (Photograph by the author.)

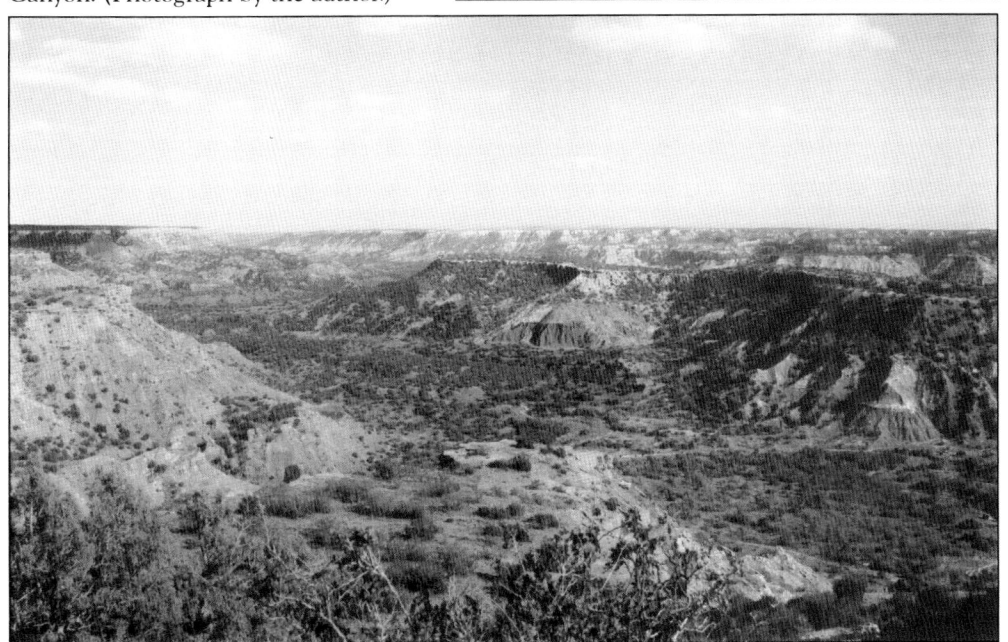

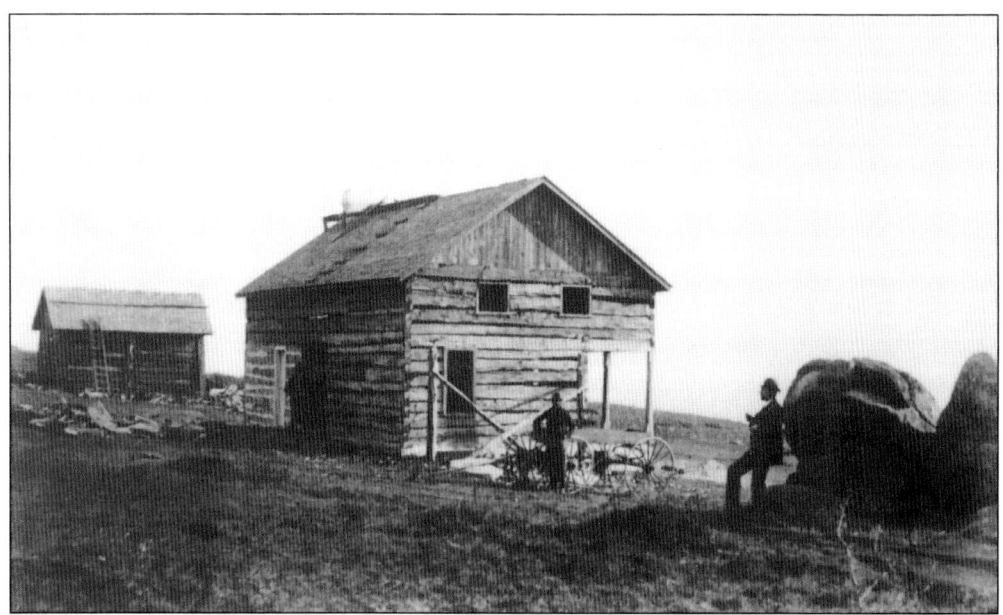

Charles Goodnight spent his first year in the recently vacated Palo Duro Canyon in a dugout. But he soon launched the development of a major ranch. Over the next decade, Goodnight supervised the construction of nearly 50 ranch houses, bunkhouses, and line cabins, along with hundreds of miles of roads and fencing, two dozen stock tanks, a tin shop, and a large blacksmith shop. He also developed a dairy, a hay farm, and a poultry yard. (Author's collection.)

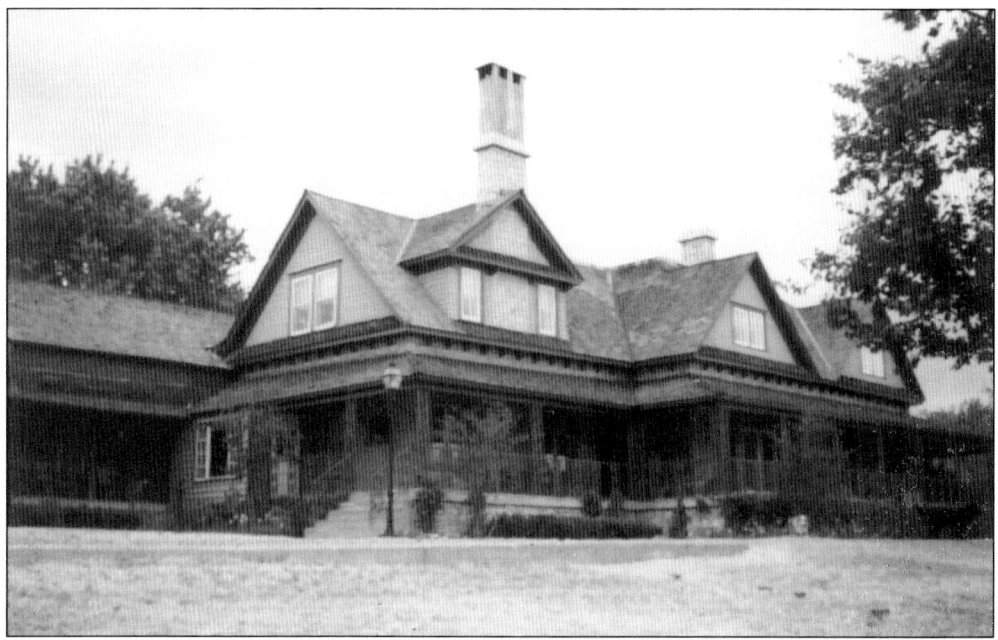

In order to develop his Palo Duro Canyon Ranch, Charles Goodnight needed financial backing. British financier John Adair provided funding for Goodnight to furnish the foundation herd and manage the ranch for $2,500 per year. Within five years, he had made profits of more than $512,000. And when John and Cordelia Adair visited the ranch, their home was this handsome stone dwelling at the headquarters complex. (Photograph by the author.)

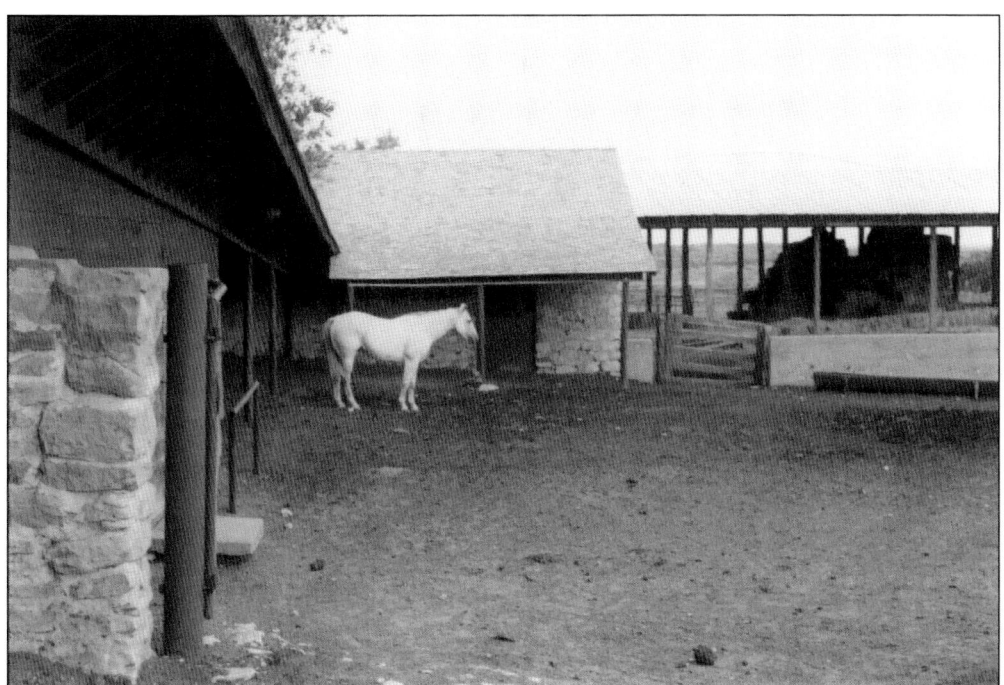

Most of the buildings at the headquarters complex still stand and are in use, including the stone horse stable. (Photograph by the author.)

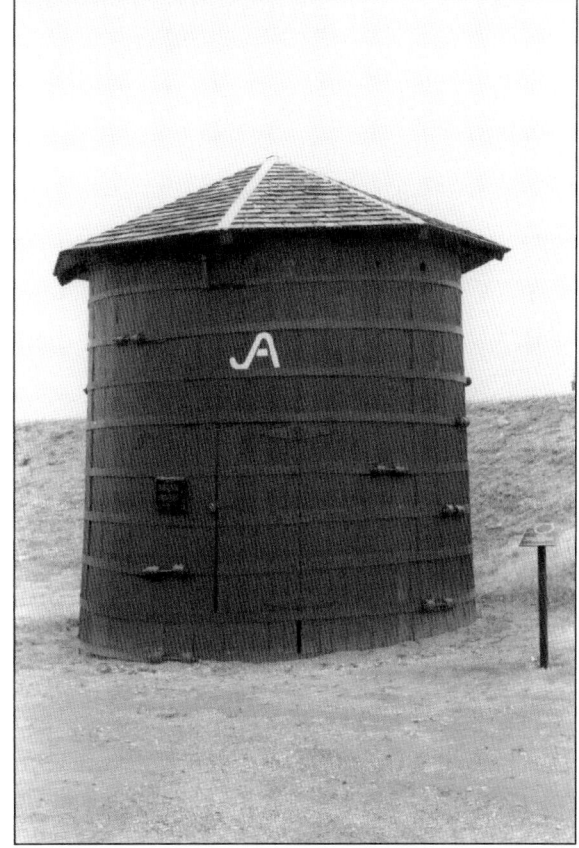

When John Adair agreed to finance Goodnight's Palo Duro venture, the spread was named the JA Ranch, but universally called the JAs. The brand, displayed on a tank built to water 3,000 JAs cattle, became famous throughout cow country. (Photograph by the author.)

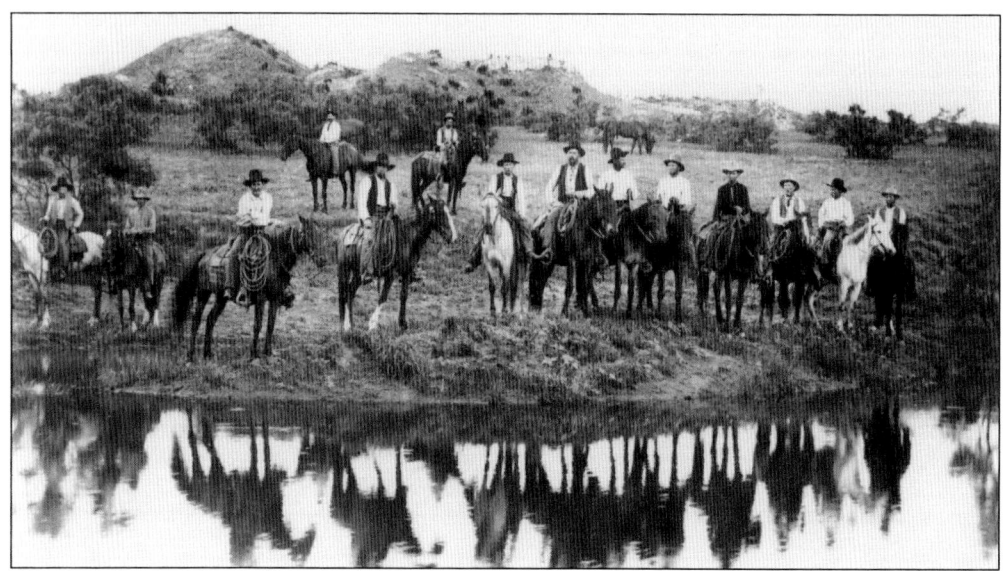

By the mid-1880s, JAs cowboys stayed busy herding more than 100,000 cattle on 1.335 million acres. Charles Goodnight was a forceful disciplinarian, imposing three rules on his riders: no drinking, no gambling, and no fighting. The steady JAs hands saved their wages—in 1885, there was $26,000 on deposit at the ranch for employees—and cowboys were allowed to run their horses and cattle on JA range. (Courtesy Panhandle Plains Historical Museum, Canyon, Texas.)

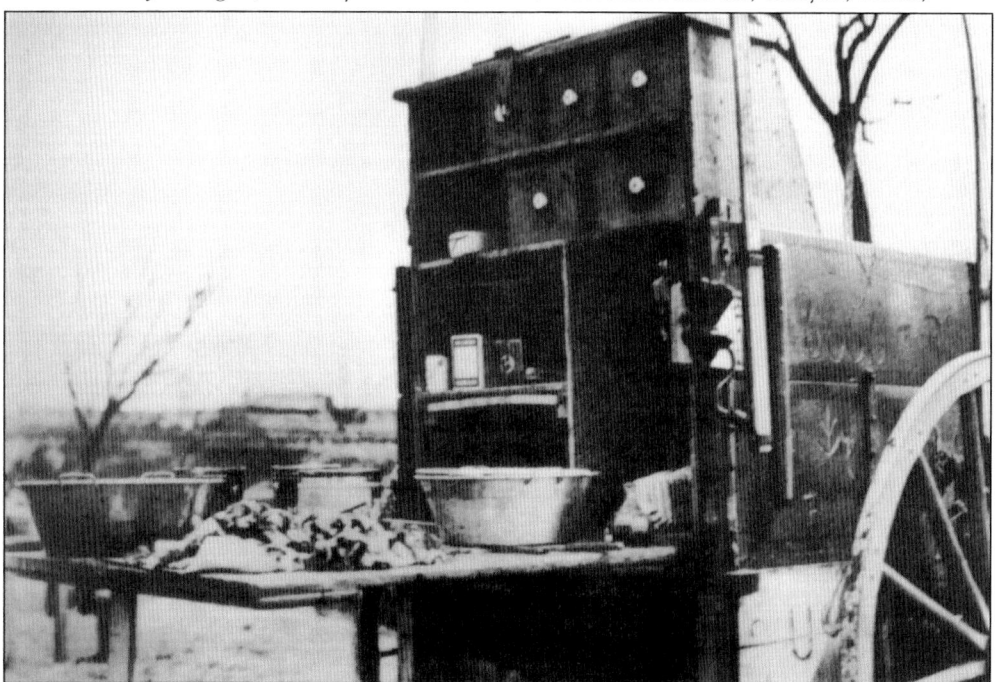

Anticipating long cattle drives northward after the Civil War, Charles Goodnight invented the chuck wagon to feed range crews. He had a military wagon rebuilt with tough bois d'arc wood and iron axles. Goodnight designed a chuck box, covered by a hinged lid that could be lowered onto a swinging leg to form the cook's worktable. Bachelors sometimes attached a chuck box to their kitchen wall as a pantry and table. (Author's collection.)

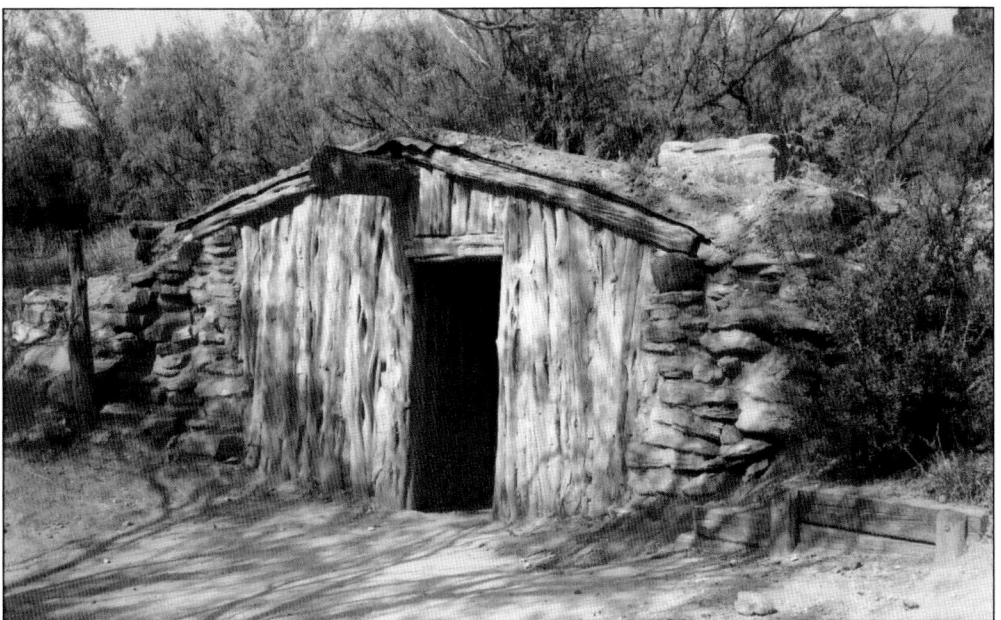

Because the JA range was so vast, during the winter months, line camps were manned at locations far from headquarters. A solitary cowboy would winter at a line cabin or dugout, riding out each day to check on JAs cattle. This dugout was built in the northern part of Palo Duro Canyon. (Photograph by the author.)

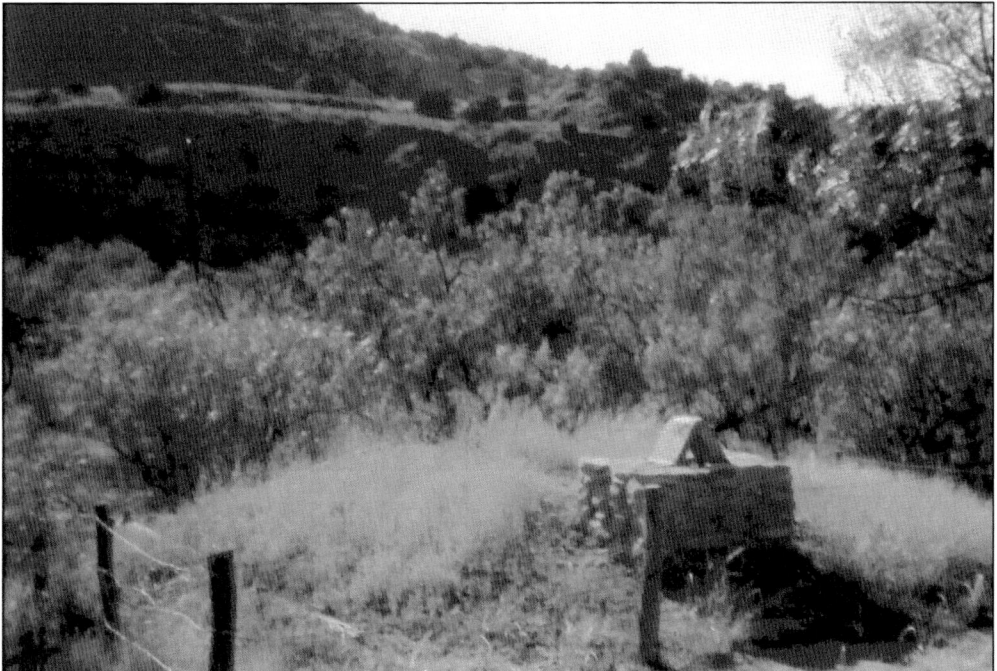

This one-room dugout had a stone fireplace and chimney in the rear for cooking and heating. Dug into a hill, only the front wall and roof had to be built. But the sod roof sprouted grass and had to be fenced off so that a grazing animal would not crash through the roof into the main room. (Photograph by the author.)

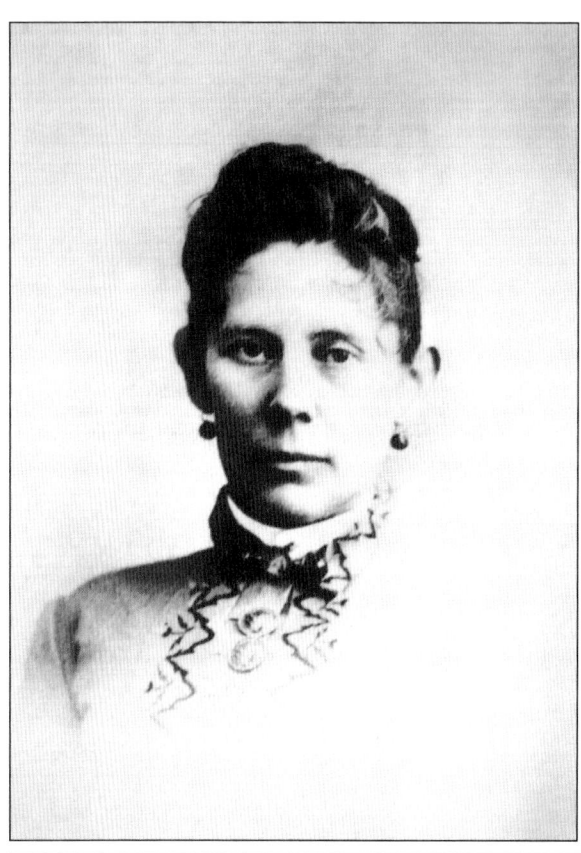

Charles Goodnight married Mary Ann "Molly" Dyer (left) in 1870, when she was 30 and he was 34. Molly accompanied her pioneer husband into Palo Duro Canyon in 1876. She made a home of their dugout and, according to her husband, "She met isolation and hardship with a cheerful heart, and danger with undaunted courage." Molly Goodnight became known as the "Mother of the Panhandle." (Courtesy Historical and Biographical Record of the Cattle Industry of Texas.)

At the JAs headquarters complex, Charles Goodnight built the comfortable stone residence below for Molly and himself. They were married for 55 years before her death in 1926. Although the Goodnights never had children, Charles testified that Molly "made her home a house of joy." (Photograph by the author.)

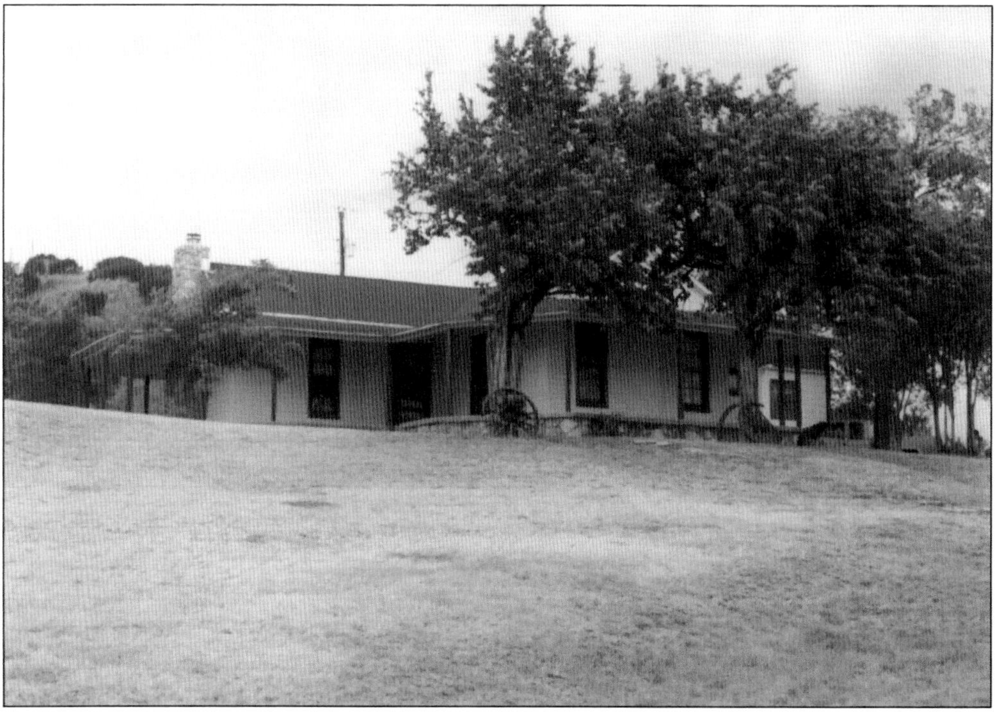

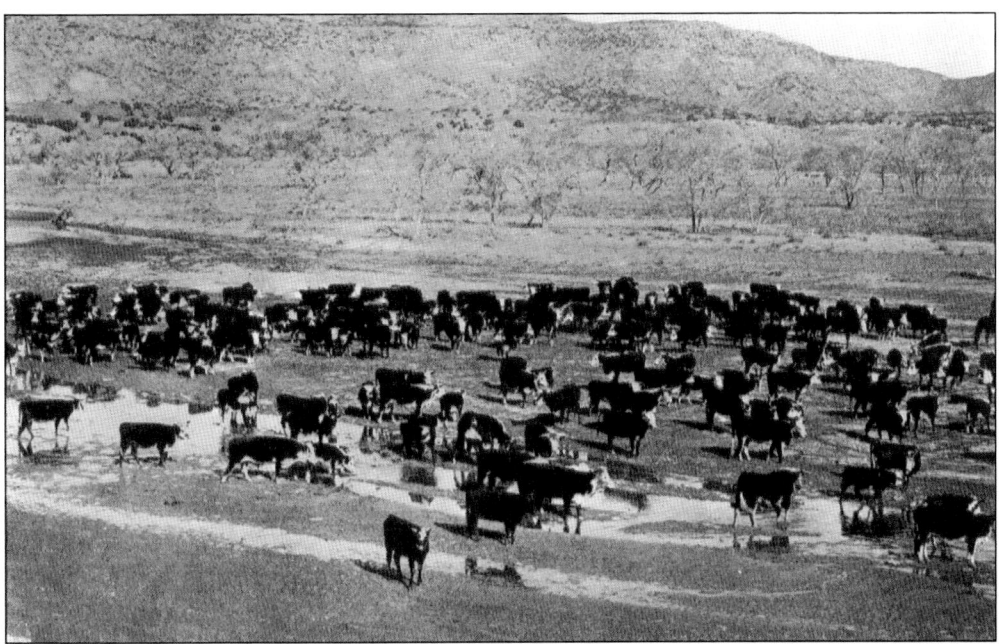

Across West Texas, ranchers began improving their beef herds with Hereford cattle. These Herefords graze on the vast JAs range in Palo Duro Canyon. (Courtesy Historical and Biographical Record of the Cattle Industry of Texas.)

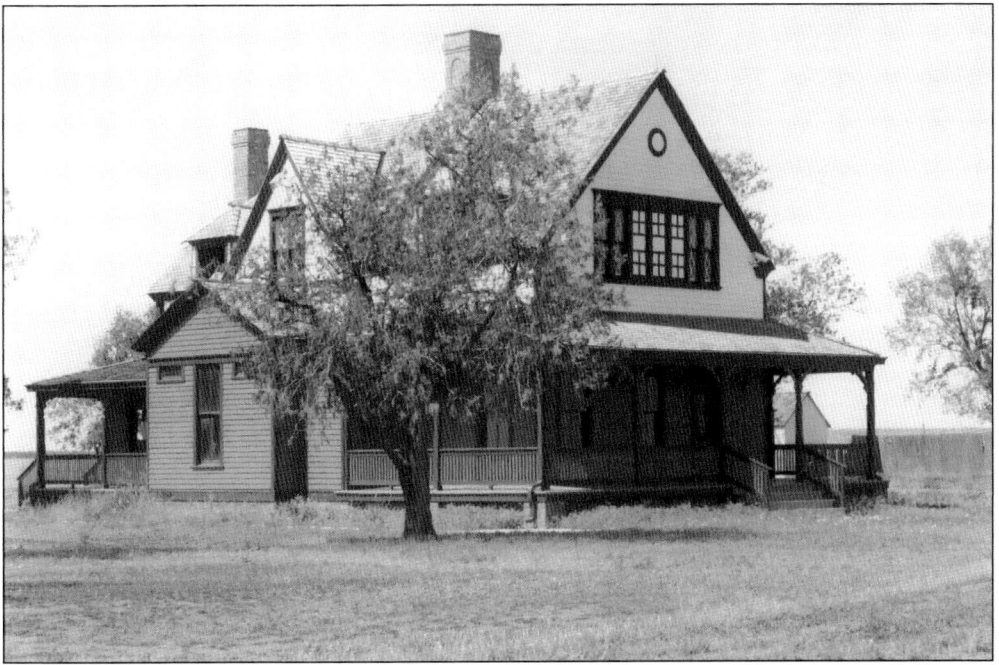

Following the death of John Adair in 1885, his ranching interests went to his wife, Cordelia, who continued to live in England. By 1887, Charles Goodnight impatiently divested himself of his interests in the JAs and moved north to a new town that was named Goodnight by the adjacent railroad. The Goodnights built this handsome Victorian home, which boasted a ballroom and a large second-floor sleeping porch. (Photograph by the author.)

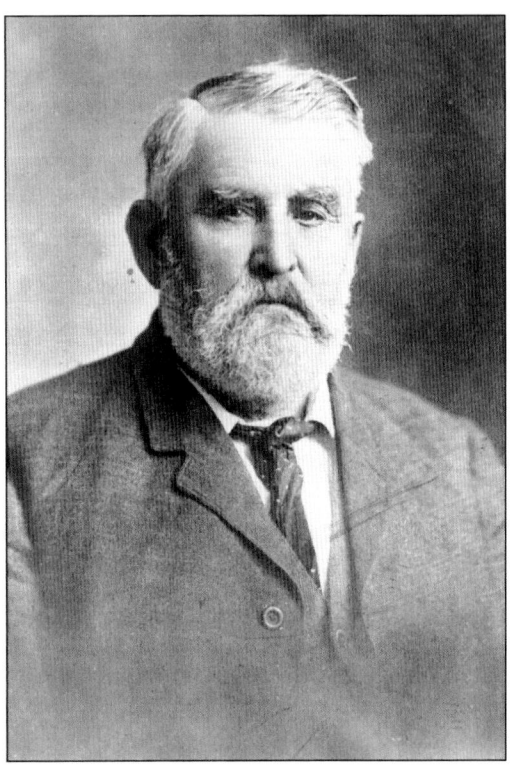

Charles and Molly Goodnight erected a school and a church in the little community that bore their name. When asked what kind of church he had supported, Goodnight replied, "I don't know, but it's a damned good one." The legendary cattleman died at 93 in 1929 and was buried beside his beloved Molly at the Goodnight Cemetery. (Courtesy of the Panhandle Plains Historical Museum, Canyon, Texas.)

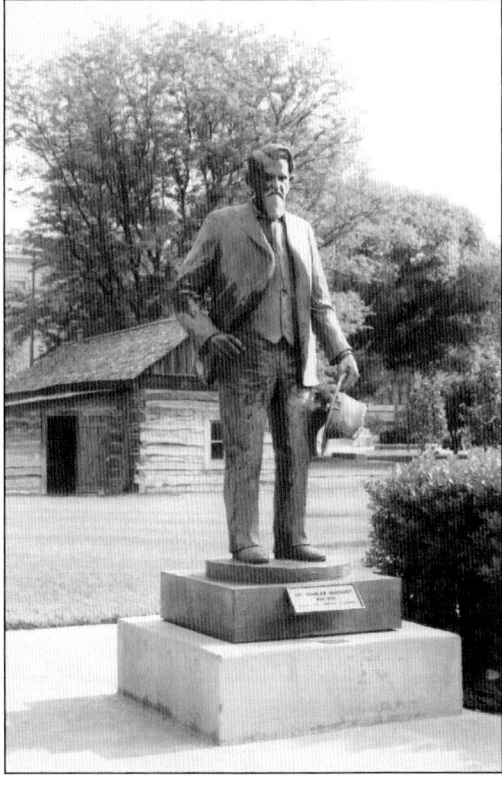

This commanding statue stands beside the Panhandle Plains Museum on the campus of West Texas State University in Canyon. Behind the statue is the headquarters cabin of the T-Anchor Ranch. (Photograph by the author.)

Four

GREAT RANCHES AND RANCHERS

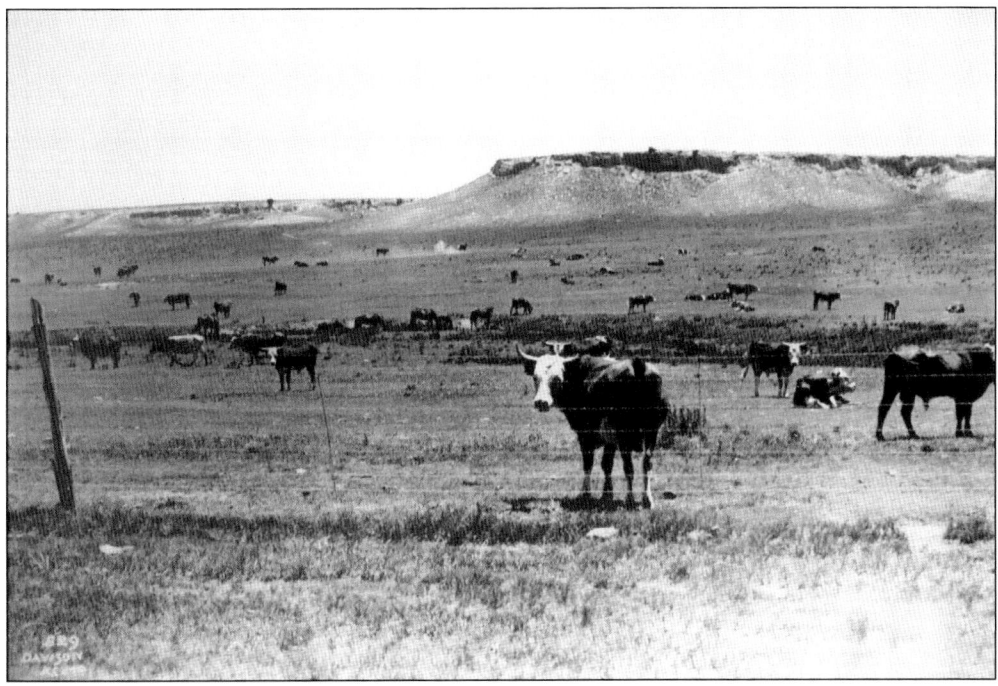

These XIT cattle graze on the largest ranch under one fence in the West. More than three million acres extended from the Texas Panhandle south for more than 200 miles along the New Mexico border, covering parts of 10 counties across the sparsely populated region. In the mid-1880s, large herds of longhorns were purchased and placed on the vast range. After 1887, there were no further purchases of major herds, although a great many Hereford, Durham, and Polled Angus bulls were brought in to improve the beef quality. By the early 1880s, the XIT had developed the largest herd of high-grade Polled Angus cattle in Texas. In its heyday, the XIT maintained herds of 125,000 to 150,000 cattle. (Courtesy Historical and Biographical Record of the Cattle Industry of Texas.)

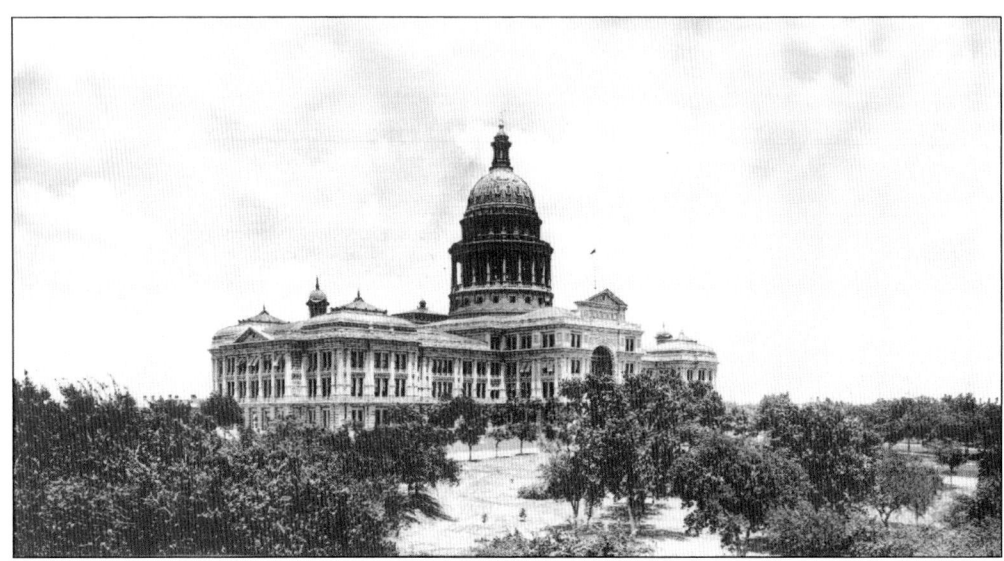

In 1879, the Texas Legislature passed a law appropriating 3.05 million acres to finance a splendid new state capitol in Austin. From this legislation emerged the Capitol Syndicate Ranch, better known as the XIT. The contract was awarded to a Chicago firm, which completed the magnificent building in 1888. (Author's collection.)

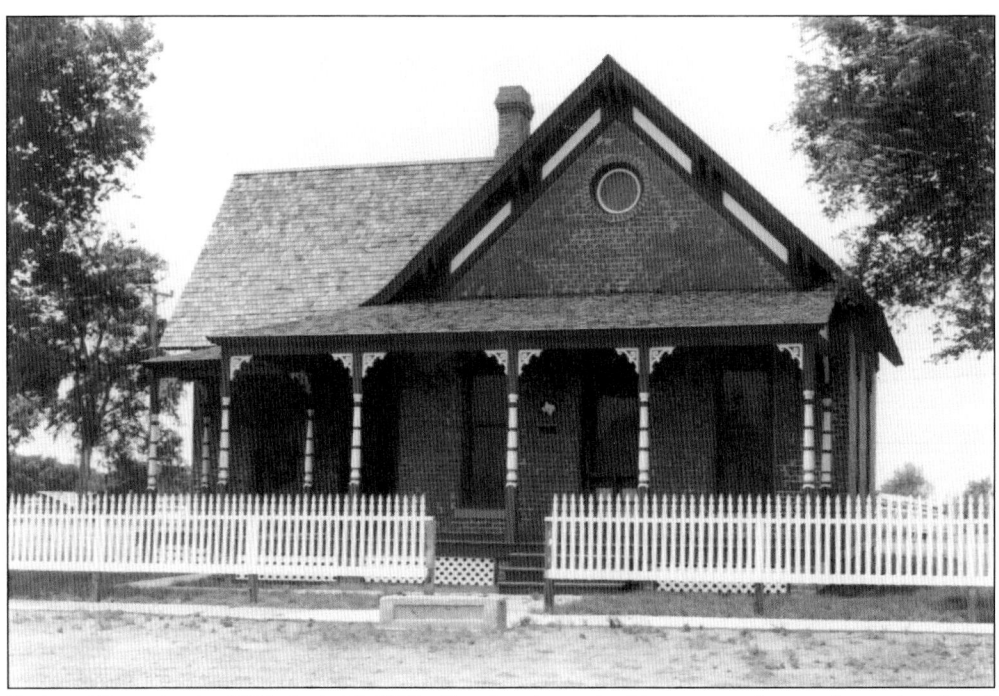

Built in Channing in 1890, the brick XIT office faced east, looking out on the north-south railroad tracks. A vault was placed inside. The XIT office stands beautifully restored today. (Photograph by the author.)

A.G. Boyce was the general manger of the XIT for nearly two decades during the height of the ranch's cattle operations. Born and raised on the Texas frontier, Boyce served as a Confederate soldier throughout the Civil War. After the war, he was active in the cattle trade and drove some of the first herds onto the new XIT. Appointed general manager in 1887, Boyce divided the vast ranch into seven divisions, directed fencing efforts, drilled 300 wells, and erected scores of buildings. (Courtesy Historical and Biographical Record of the Cattle Industry of Texas.)

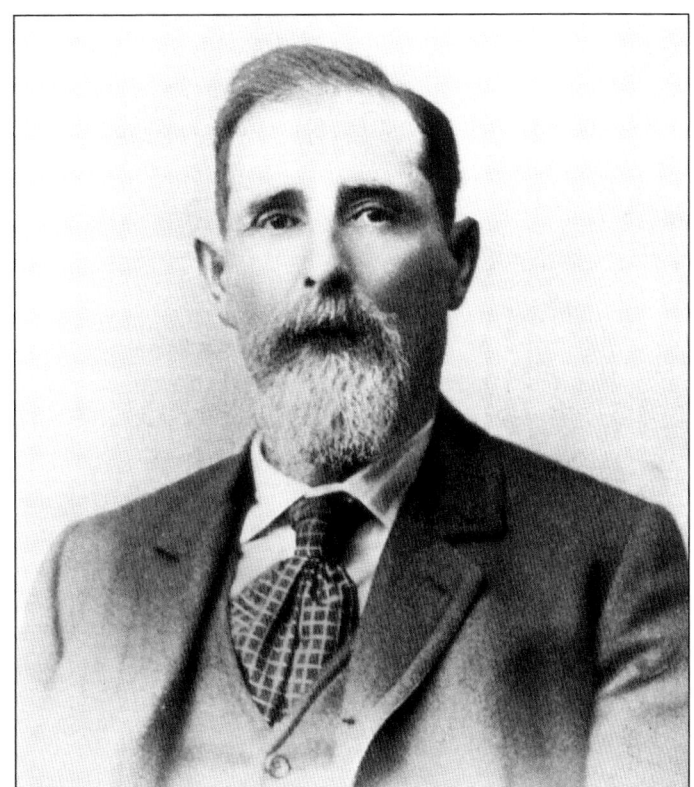

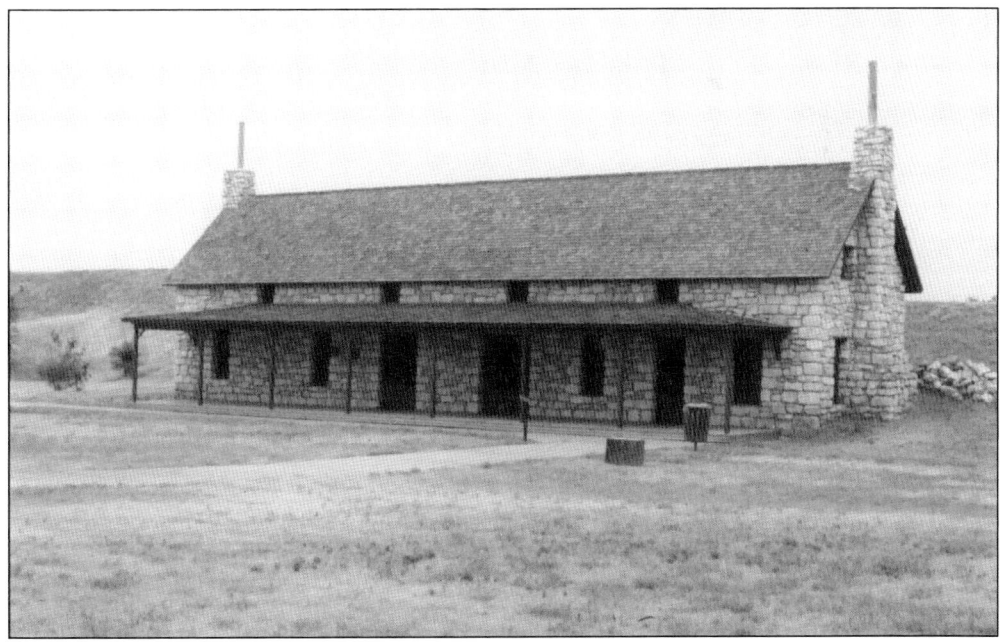

One of the seven divisions of XIT was the Las Escarbadas Division, a breeding range that included this two-story stone bunkhouse, which was built in 1886 near the New Mexico border. The upper level was filled with cots. At floor level, the left half was a kitchen and a large mess hall and the right half was foreman's quarters. (Photograph by the author.)

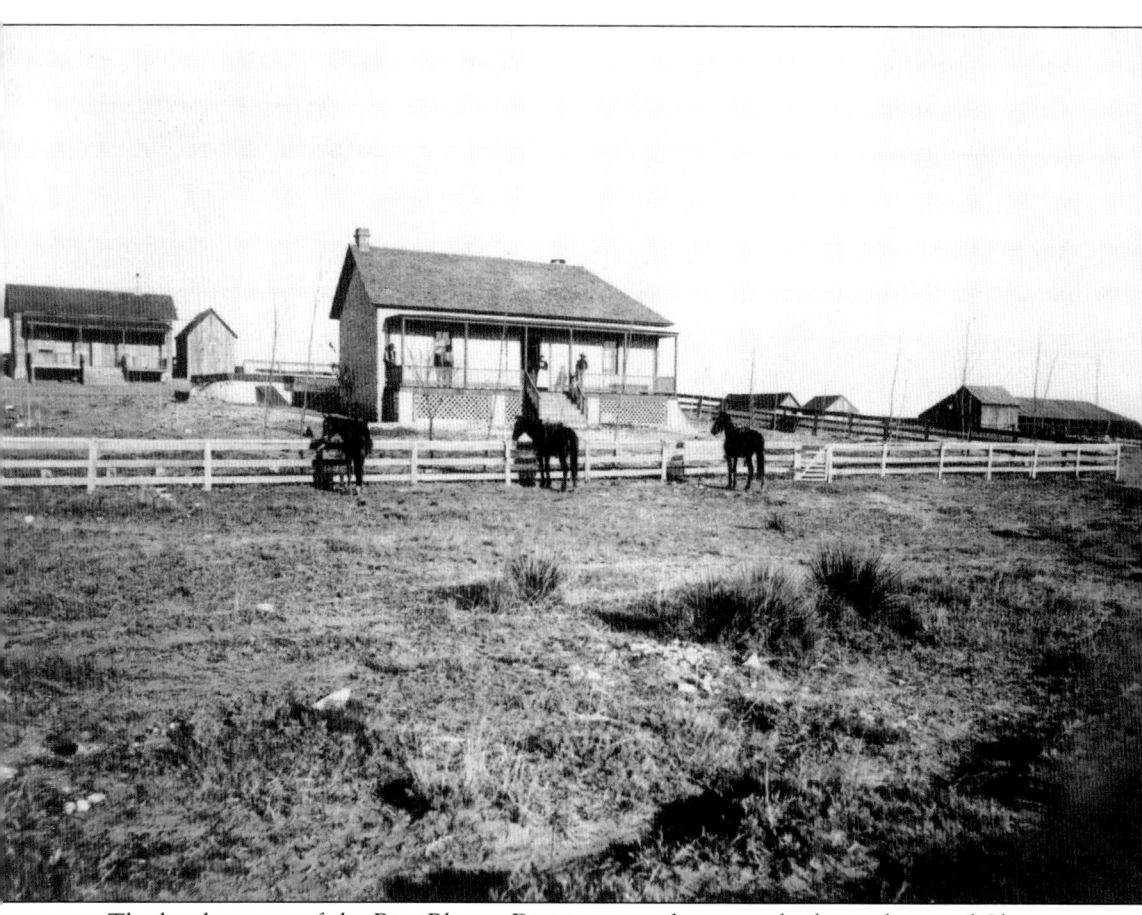

The headquarters of the Rita Blanco Division, seen here, was built northwest of Channing. From north to south, the divisions were: Buffalo Springs, Middle Water, Ojo Bravo, Rita Blanco, Las Escarbadas, Spring Lake, and Yellow House. Fencing operations in the 1880s brought 300 carloads of materials to the XIT. The great ranch was divided into 94 pastures, requiring about 1,500 miles of fence. Some 6,000 miles of barbed wire was used, along with 100,000 cedar posts, five carloads of wire staves, an entire carload of staples, and a carload of hinges for the hundreds of gates. Line riders kept a constant check on fences, and some divisions kept fence wagons in operation at all times. (Courtesy Panhandle-Plains Museum, Canyon.)

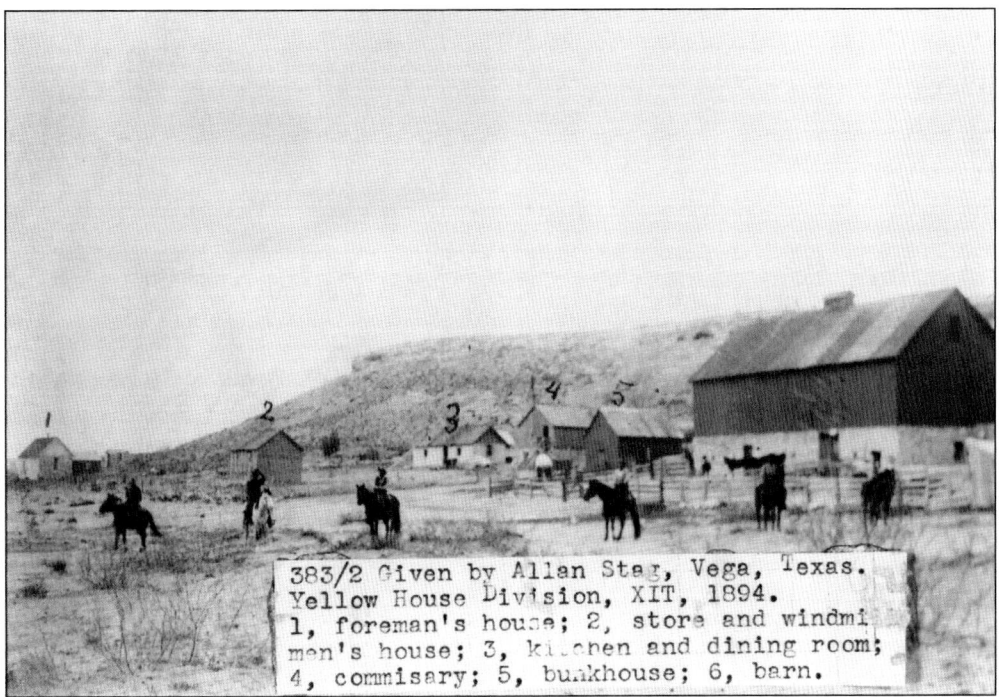

The southernmost division of the XIT was the Yellow House, whose headquarters complex is seen here in 1894. Most of these buildings remain in use today. (Courtesy Panhandle-Plains Museum, Canyon.)

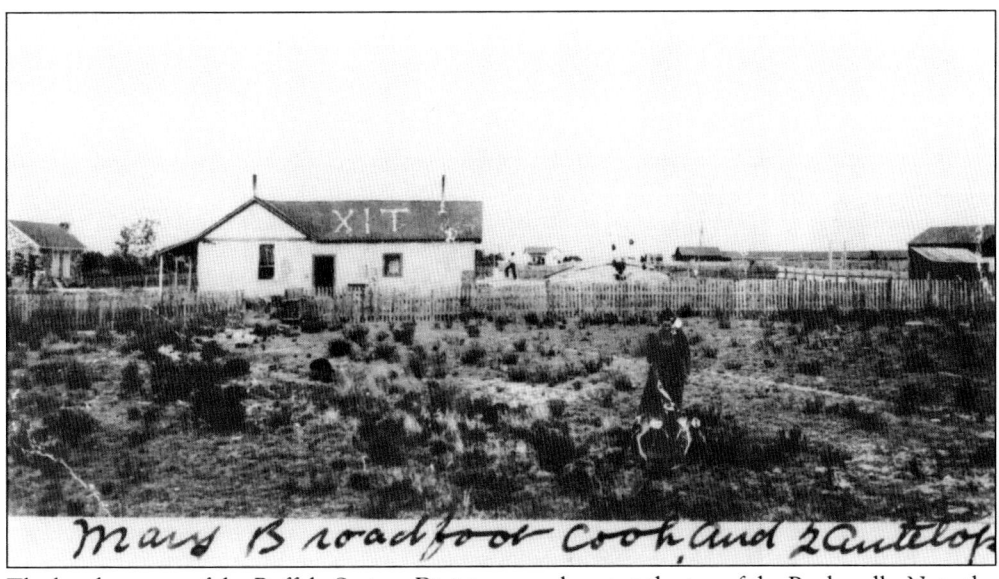

The headquarters of the Buffalo Springs Division was almost at the top of the Panhandle. Note the "XIT" painted on the roof of the ranch house. In the foreground is the cook, Mary Broadfoot, with two baby antelopes. (Courtesy Dobie Collection, Academic Center, University of Texas at Austin.)

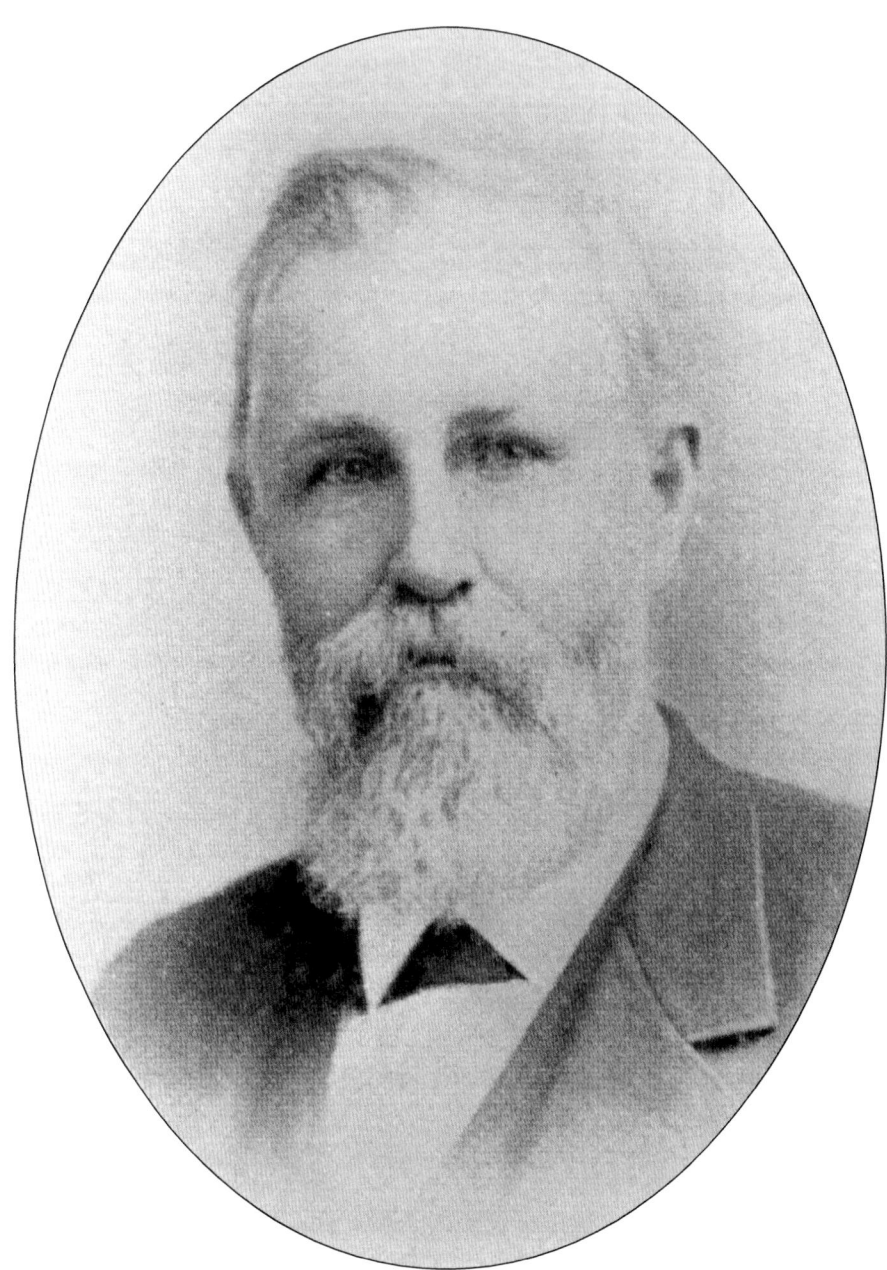

"I am Shanghai Pierce, Webster on cattle, by God, sir." The big flamboyant rancher who boomed out this announcement to a hotel clerk was, indeed, cattleman enough to become legendary throughout the West. "I stand pat I am the best cowman in Texas," he boasted. A strapping six foot, four inches, Shanghai came to Texas from his native Rhode Island in 1853 and went to work on a Matagorda County ranch, taking his pay in cattle. His real name was Abel Pierce, and he branded AP on any unbranded animal he encountered. In 1865, he and his younger brother, Jonathan, founded El Rancho Grande, soon branding 18,000 calves in a single year. Constantly riding to form partnerships and buy more cattle, Shanghai later bragged, "I owned nearly all the cattle in Christendom once." (Courtesy Historical and Biographical Record of the Cattle Industry of Texas.)

John Chisum was reared on a Tennessee plantation, where he displayed such an affinity for cattle that he was nicknamed "Cow John." In 1837, when he was 13, he moved with his family to the fledgling community of Paris in north Texas. Chisum established a ranch north of Fort Worth in 1854. During the Civil War, he supplied beef to the Confederate army, and after the war, he moved his cattle operation to West Texas. In 1867, he ventured to New Mexico Territory, soon running 80,000 cattle for 150 miles along the Pecos River. Known as the "King of the Pecos," his 100 riders included Billy the Kid. Chisum died at 60 in 1884 and was buried in a family cemetery at Paris (below). (Right, courtesy Historical and Biographical Record of the Cattle Industry of Texas; below, photograph by the author.)

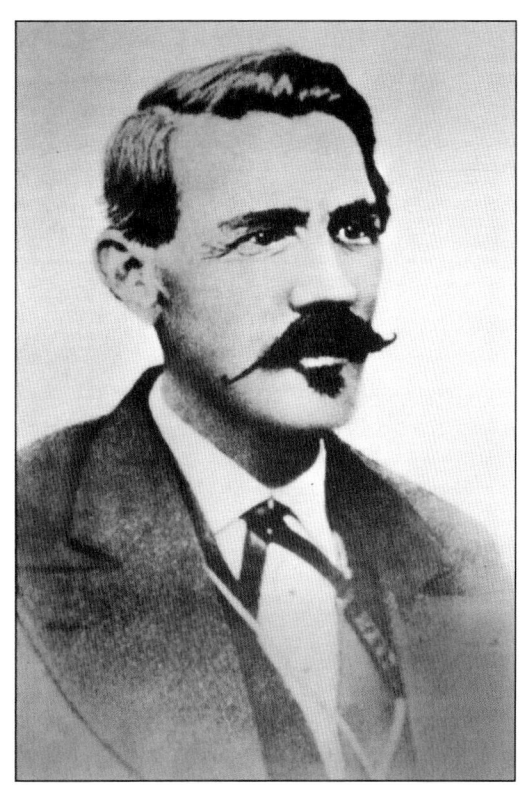

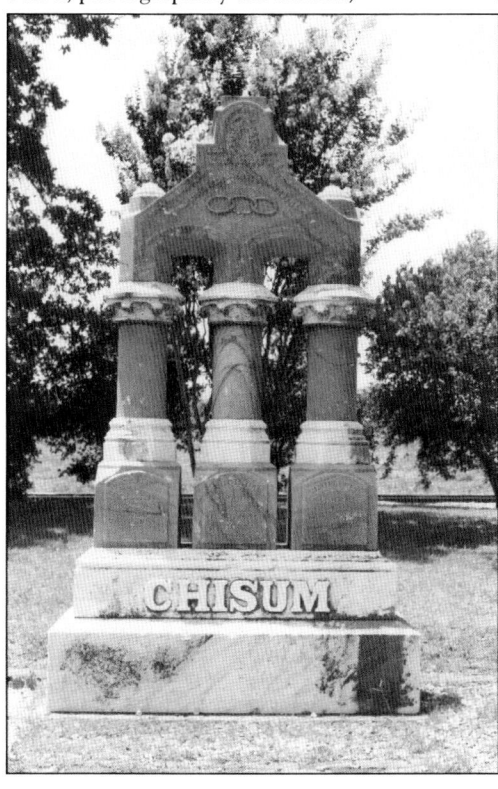

Born in Scotland in 1850, Murdo Mackenzie acquired experience in law and banking. At the age of 35, he came to Colorado to manage a cattle company financed from Edinburgh. Five years later, he agreed to manage the enormous Matador Ranch, founded in 1879 in northwest Texas. Mackenzie ran the Matador for 36 years, traveling the pastures in a horse-drawn buggy. (Courtesy Panhandle-Plains Museum, Canyon.)

The Matador was owned by a Scottish syndicate based in Dundee. The syndicate provided sound leadership for seven decades, and there were enough regular visits from Scottish management that the original office (below) was known as the "Scotch dive." (Photograph by the author.)

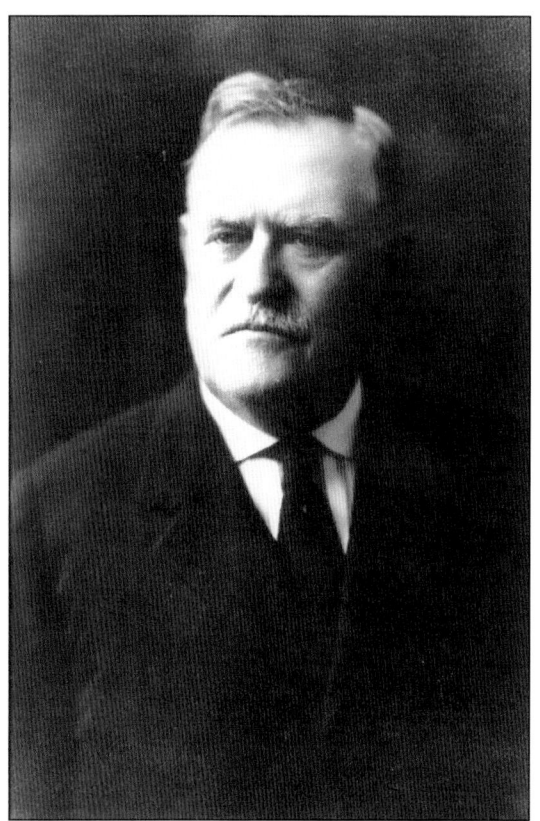

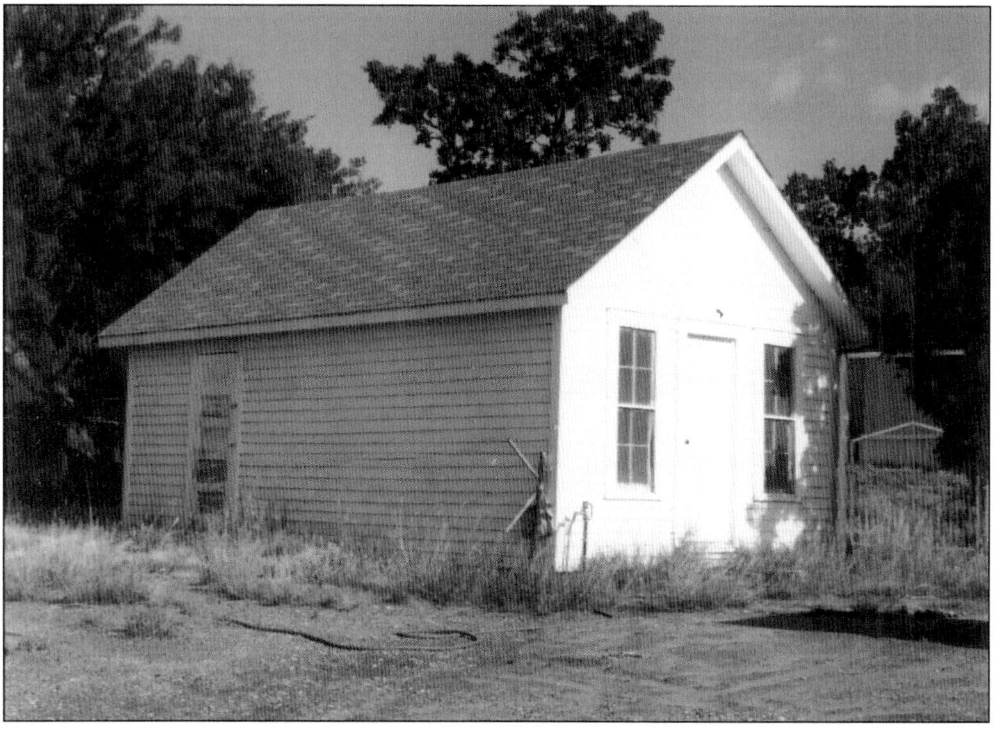

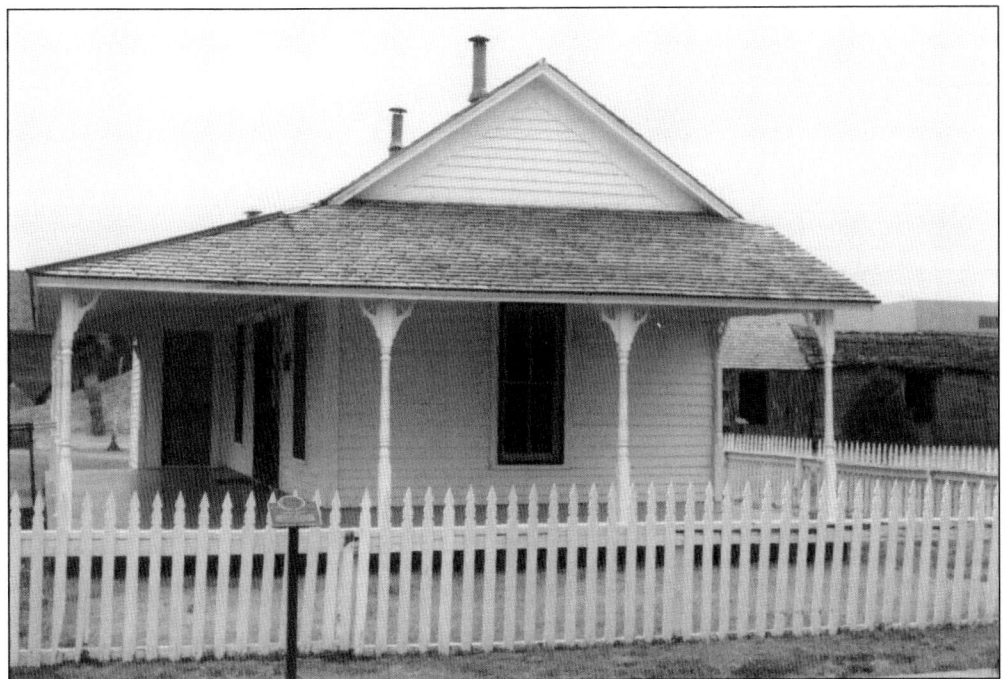

Built about 1880 as a guesthouse for its absentee Scottish owners, this frame structure eventually became the business office for the Matador Land and Cattle Company. Today, it is part of the Ranch Heritage Center in Lubbock. (Photograph by the author.)

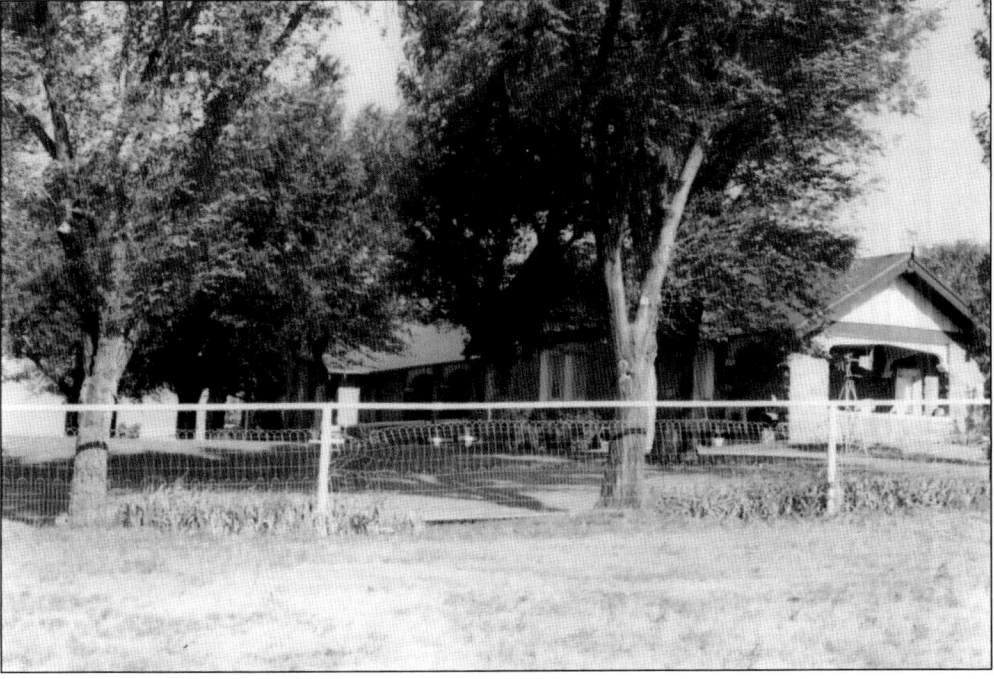

The main house at the Matador was built atop a hill just south of the town that was named for the ranch. Like most of the venerable headquarters buildings, it is still in use. (Photograph by the author.)

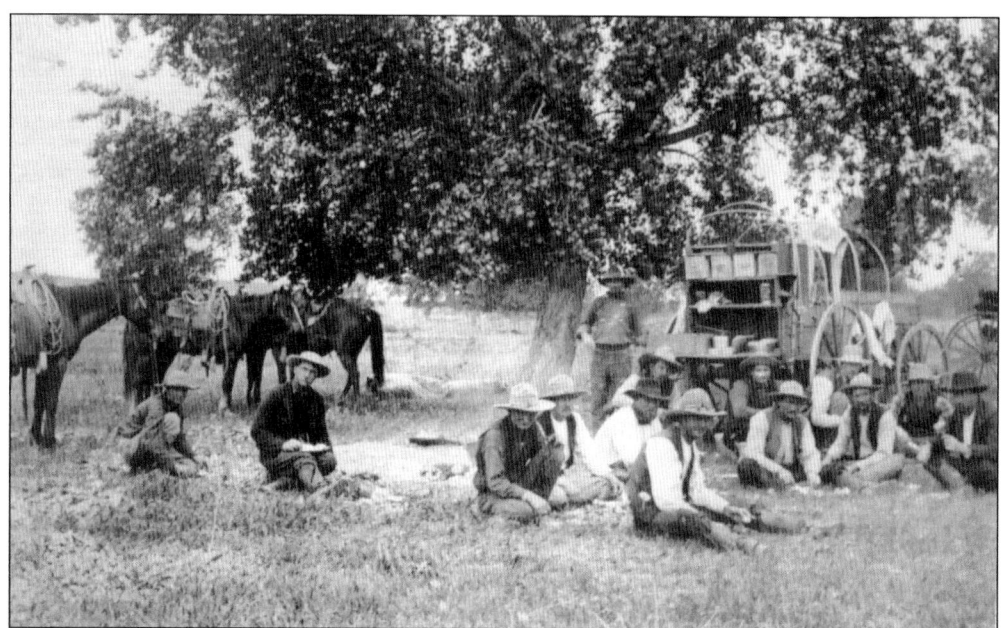

Three dozen Matador cowboys worked through the winter months each year, but the work force expanded to at least 75 during roundup seasons. The Matador range was broken and brushy, and 70,000 half-wild cattle grazed the 800,000 rugged acres. This Matador crew breaking at the chuck wagon had their work cut out for them. (Author's collection.)

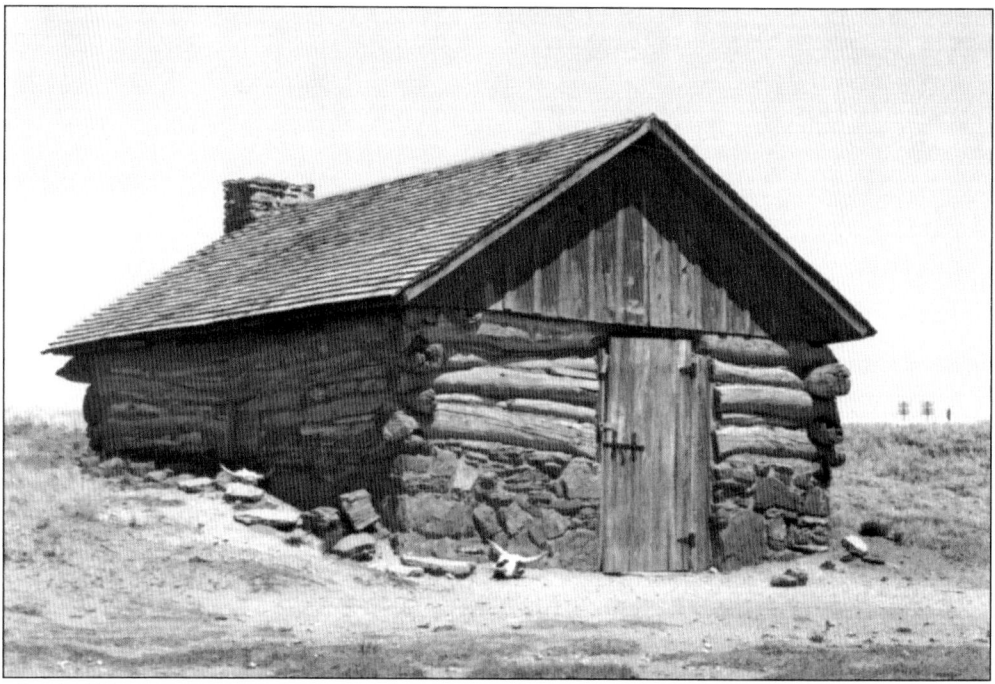

This half-dugout was built as a Matador line cabin about 1890. A line rider spent lonely winter months checking fencing and cattle on outlying ranges. Sometimes, two line riders spent the winter together, riding in different directions during the day, but providing each other companionship when off-duty. (Courtesy Ranch Heritage Center, Lubbock.)

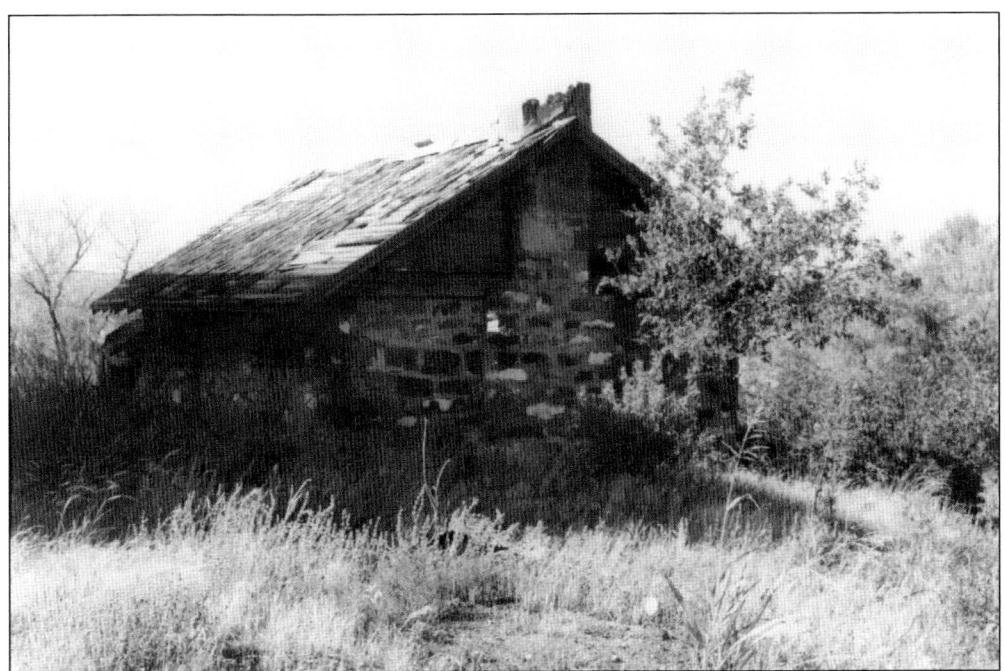

One of the earliest structures built at Matador headquarters was this stone cabin. As the headquarters complex expanded, two employees lived here and were assigned to keep the buildings tidy. (Photograph by the author.)

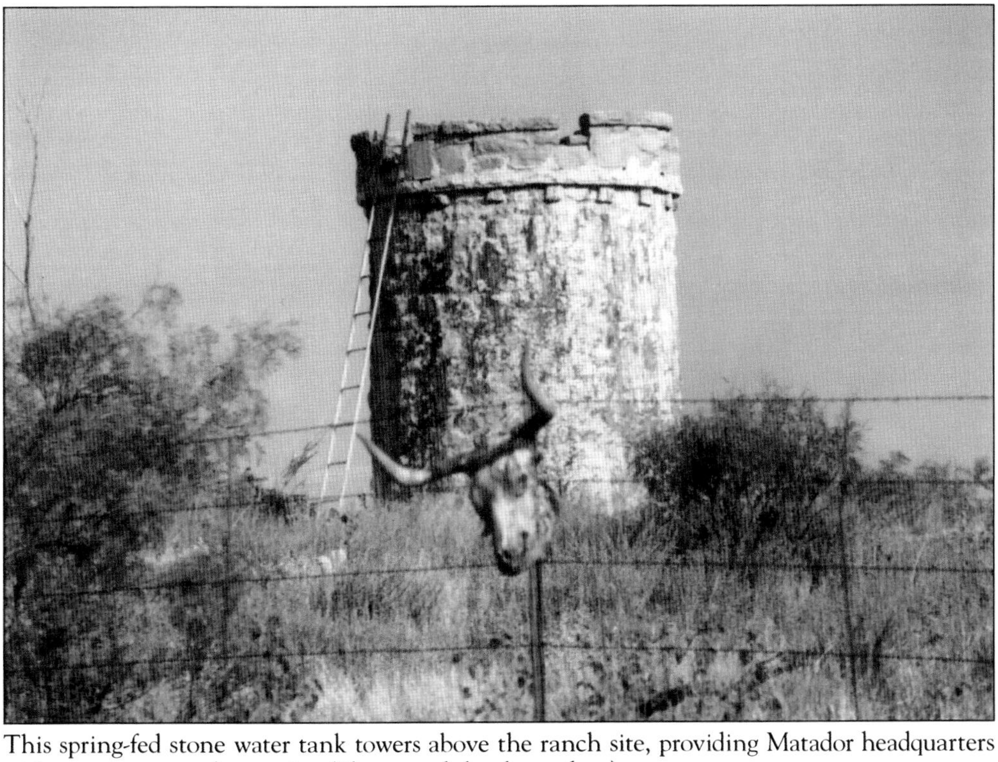

This spring-fed stone water tank towers above the ranch site, providing Matador headquarters with running water by gravity. (Photograph by the author.)

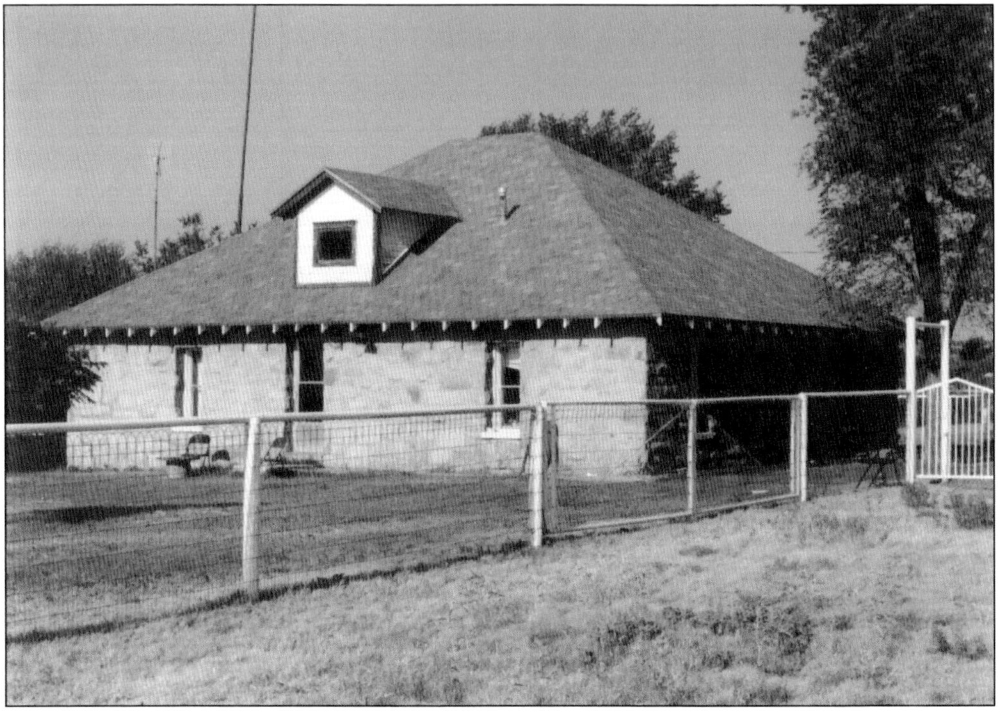

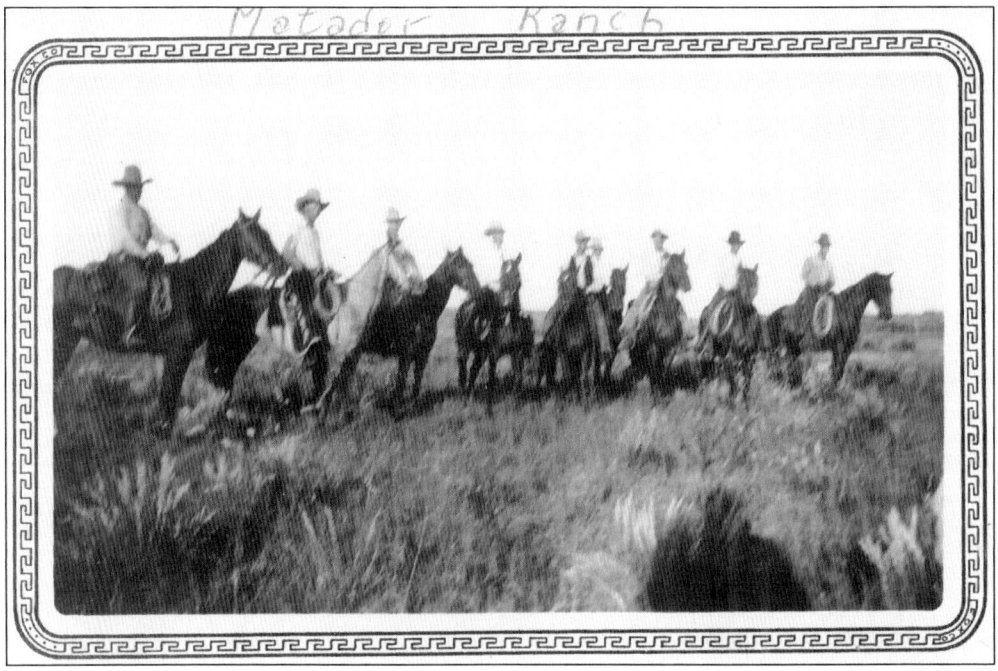

Built about 1916 as the mess house, this stone structure is now an employee residence. (Photograph by the author.)

Herb Collins, mounted on the far right, was part of a group of Matador cowboys in 1924. (Courtesy Herb Collins.)

As a young man, Henry B. Sanborn lived with Joseph F. Glidden in Illinois before going west to engage in ranching. After Glidden patented the first practicable barb wire in 1874, Sanborn aggressively sold the product in Texas. His ranching interests included the Frying Pan, and he was instrumental in founding Amarillo. (Author's collection.)

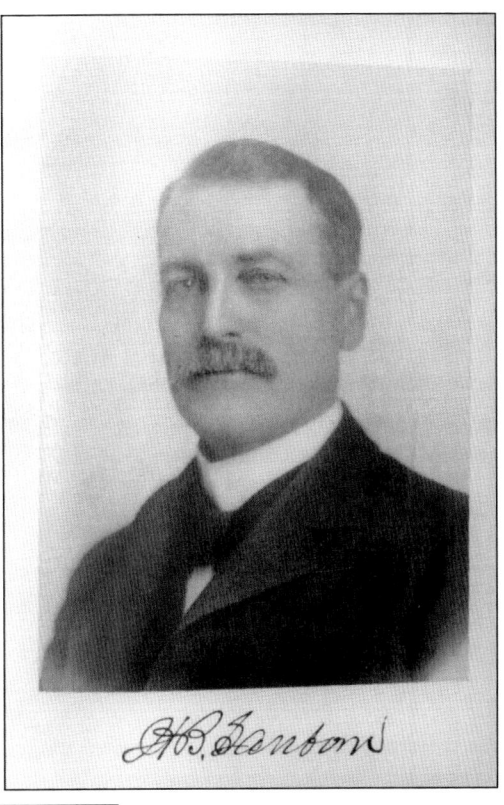

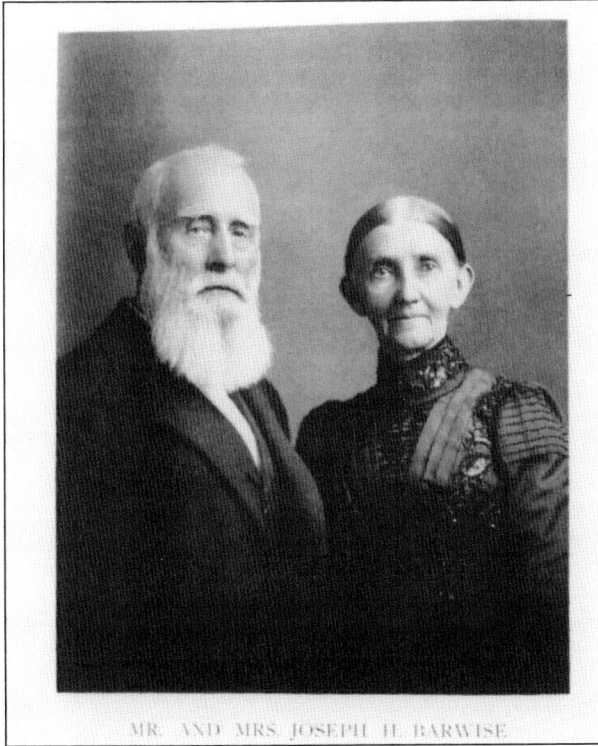

Wichita County judge J.H. Barwise, seen here with his wife, Lucy, helped found Wichita Falls. For his ranch, Bar Ys, Judge Barwise devised a brand that uniquely and cleverly expressed his last name as a horizontal bar with two "Ys" beneath it. (Author's collection.)

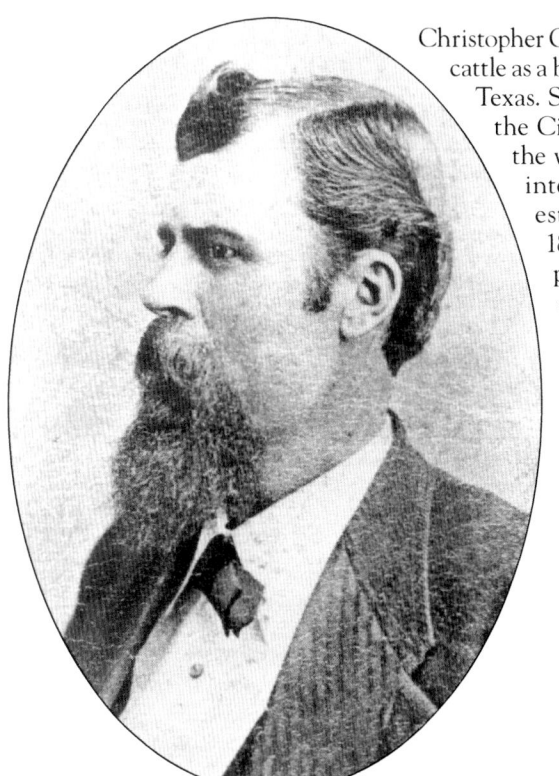

Christopher Columbus "C.C." Slaughter began handling cattle as a boy in the 1840s on the frontier of northwest Texas. Slaughter helped battle Comanches during the Civil War and drove cattle to market after the war. In addition to his successful ranching interests, he engaged in banking and real estate. Although he moved to Dallas in 1873, ranching in West Texas remained his predominant activity. (Courtesy Historical and Biographical Record of the Cattle Industry of Texas.)

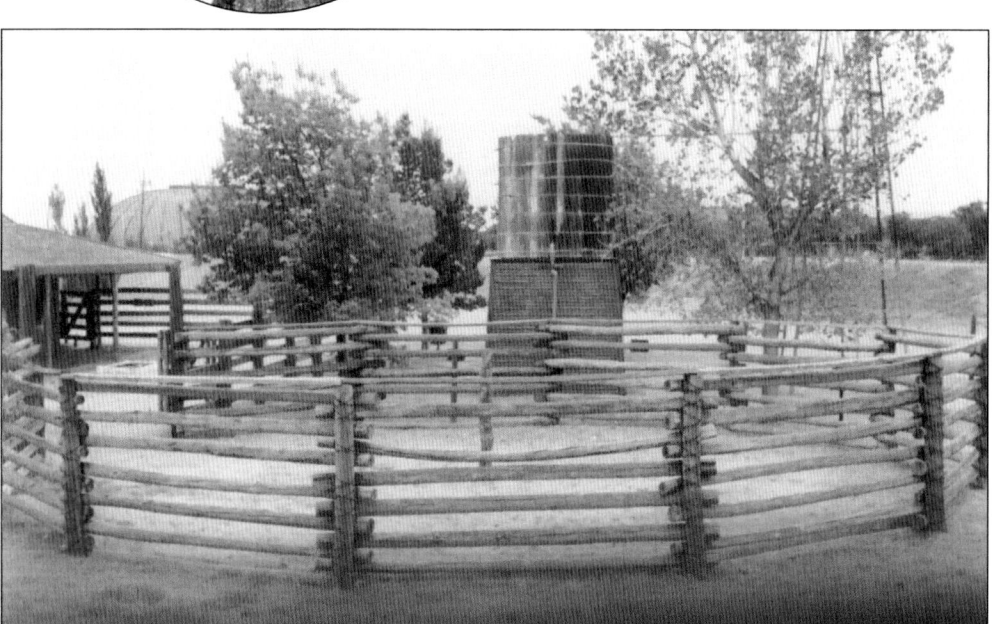

By 1878, C.C. Slaughter had formed his enormous Long S Ranch in West Texas. "Slaughter country," as it was known, stretched for 50 miles along the Colorado River, and he controlled nearly one million acres of rangeland. Rank horses were tied to the snubbing post in the middle of this Slaughter corral; the circular shape with no sharp corners protected horses and riders. (Photograph by the author.)

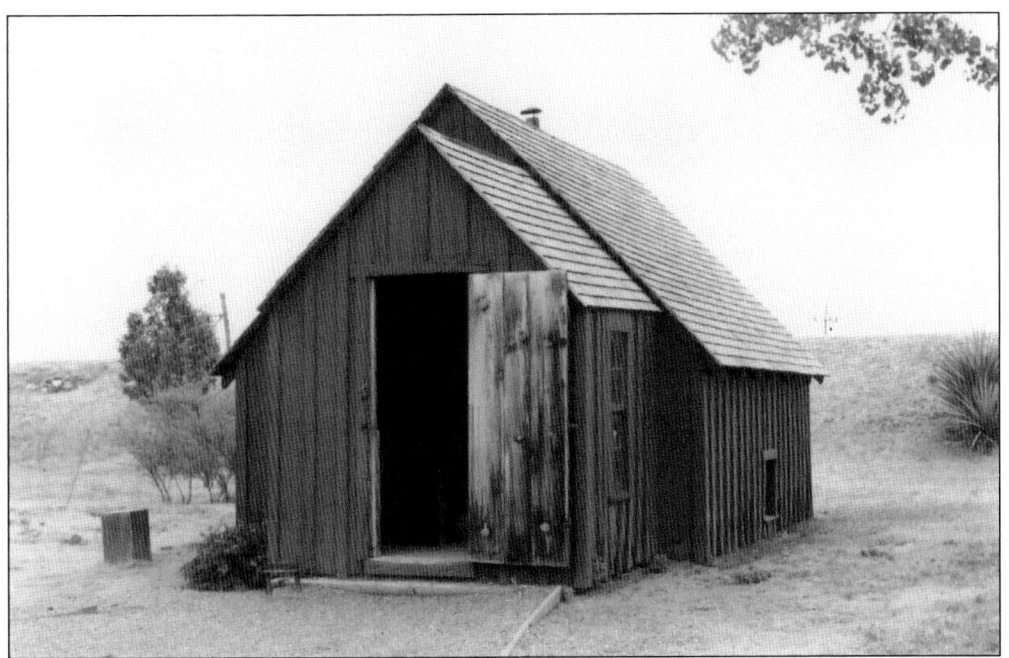

In the late 1890s, C.C. Slaughter decided to increase his West Texas holdings with another large ranch, the Lazy S. With both the Long S and the Lazy S, he owned 1.4 million acres that were stocked with 54,500 head of cattle. The first headquarters of the Lazy S was a half-dugout. This two-level dugout, with the kitchen on the lower level and the bedroom above, was the headquarters of the Lazy S Whiteface camp. (Photograph by the author.)

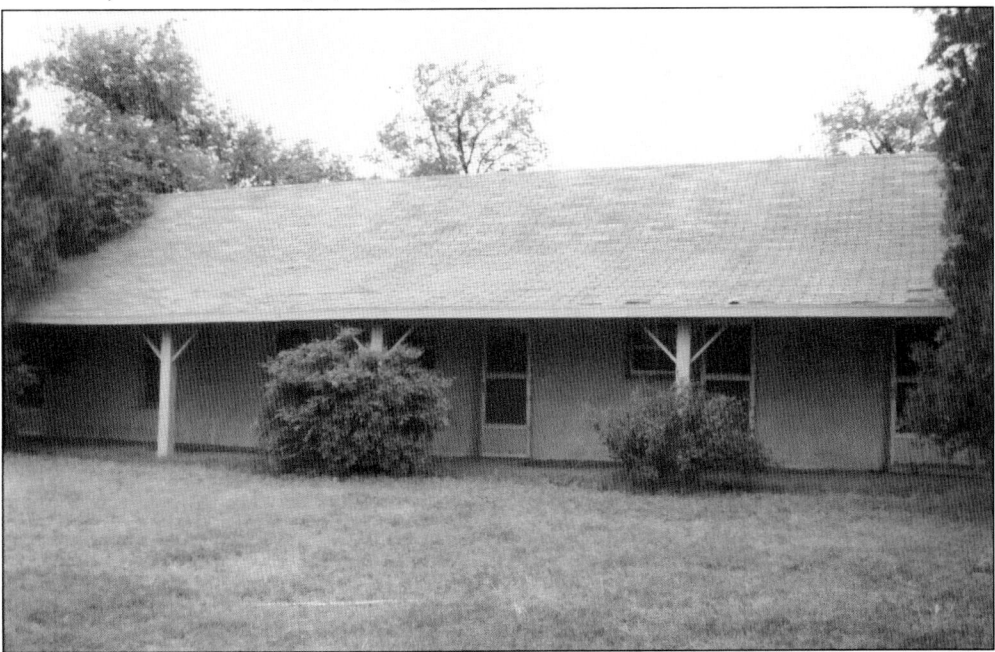

In 1915, Slaughter had a headquarters compound built for the Lazy S Ranch. On the north side of the quadrangle was the main house, 80 feet long and 25 feet wide. The bunkhouse, which faces the main house, is similar in size. (Photograph by the author.)

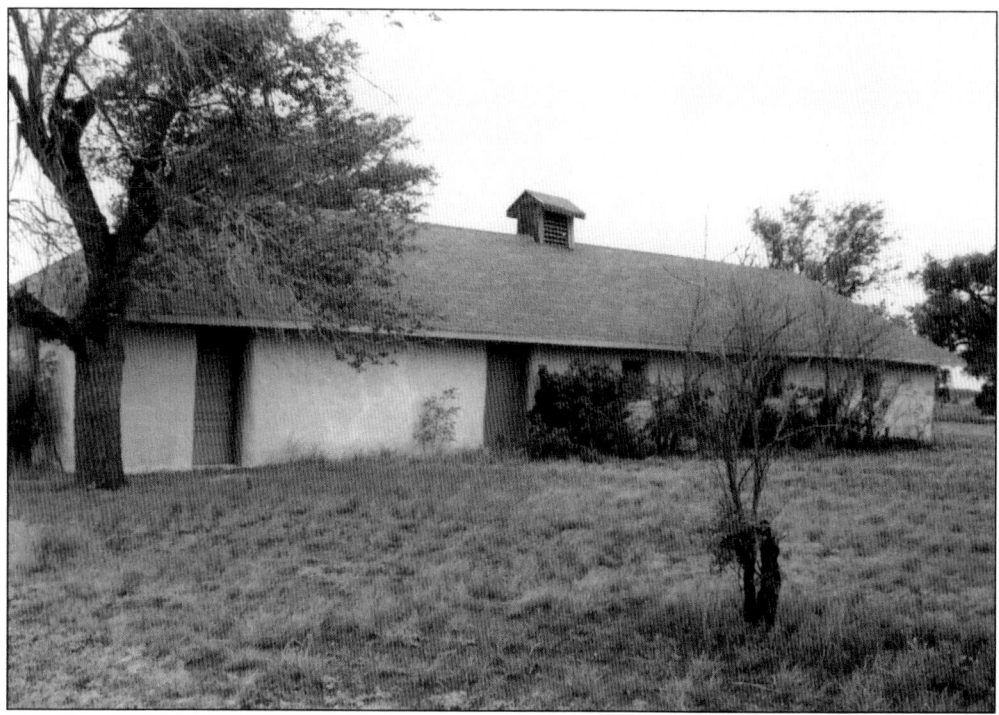

The stable, 75 feet long, was built on the west side of the Lazy S compound and was originally flat-roofed. Mexican craftsmen were brought in to build the adobe and concrete structures. (Photograph by the author.)

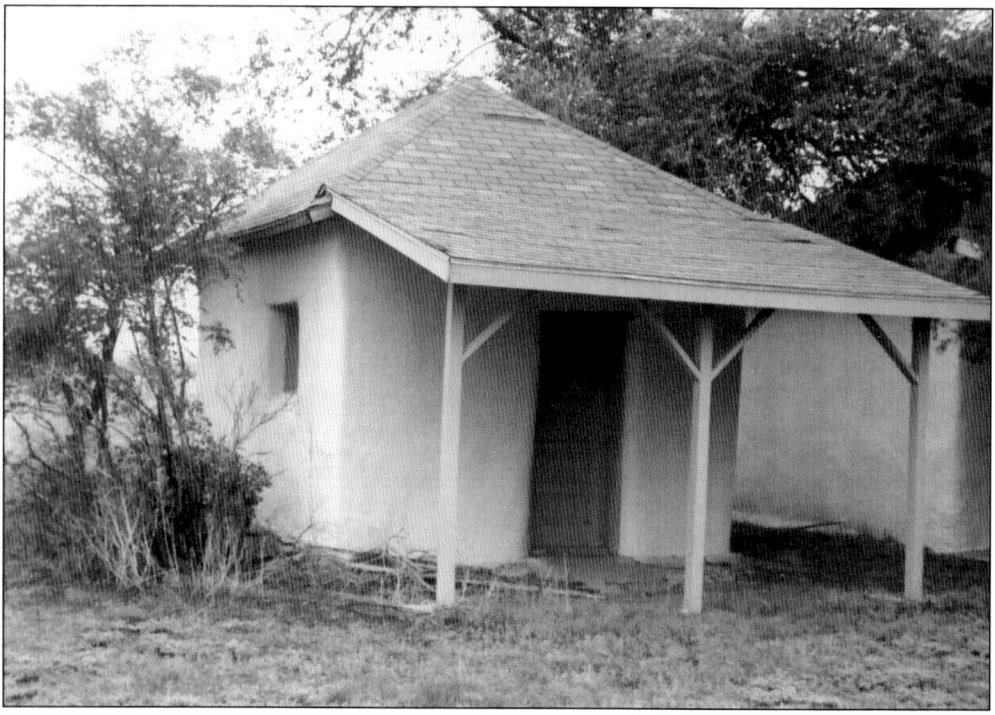

The milk house was built beside the main house. (Photograph by the author.)

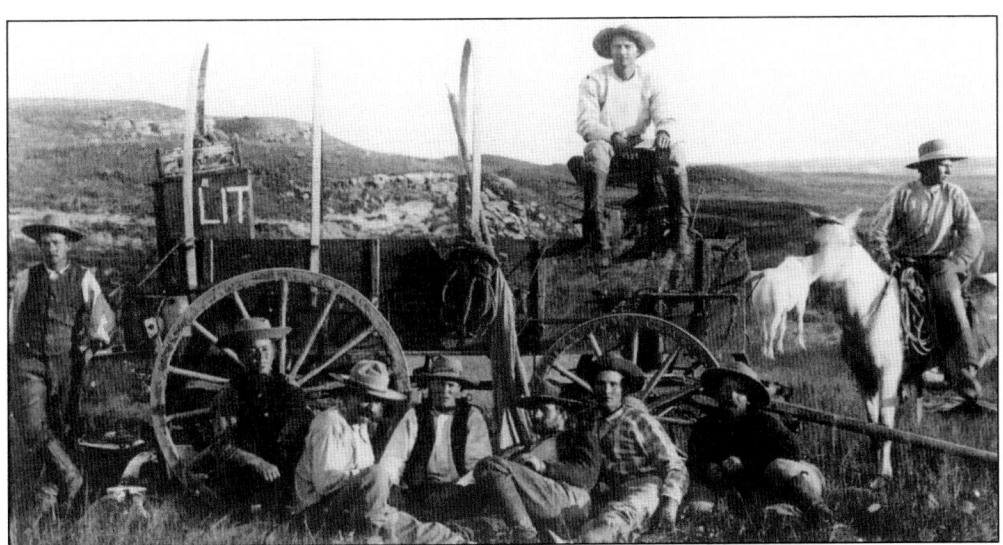

The chuck wagon above carries the brand of the LIT Ranch, established in 1877 by Maj. George W. Littlefield. Littlefield was the youngest company commander in Terry's Texas Rangers during the Civil War, rising to the rank of major before a severe leg wound sent him home to Texas. In 1871, Littlefield drove a herd of cattle to Abilene, and great profits led to more trail drives. In 1877, he bought water rights along the Canadian River near Tascosa and founded the Littlefield Ranch. He sold the LIT in 1881 for $248,000 but continued to engage in large-scale ranching in West Texas and New Mexico, branding cattle with "LFD." Littlefield moved to Austin in 1883, founding the American National Bank and serving on the board of trustees of the University of Texas. His magnificent Victorian home (below) was built just north of campus and was donated to the university. (Both, courtesy Historical and Biographical Record of the Cattle Industry of Texas.)

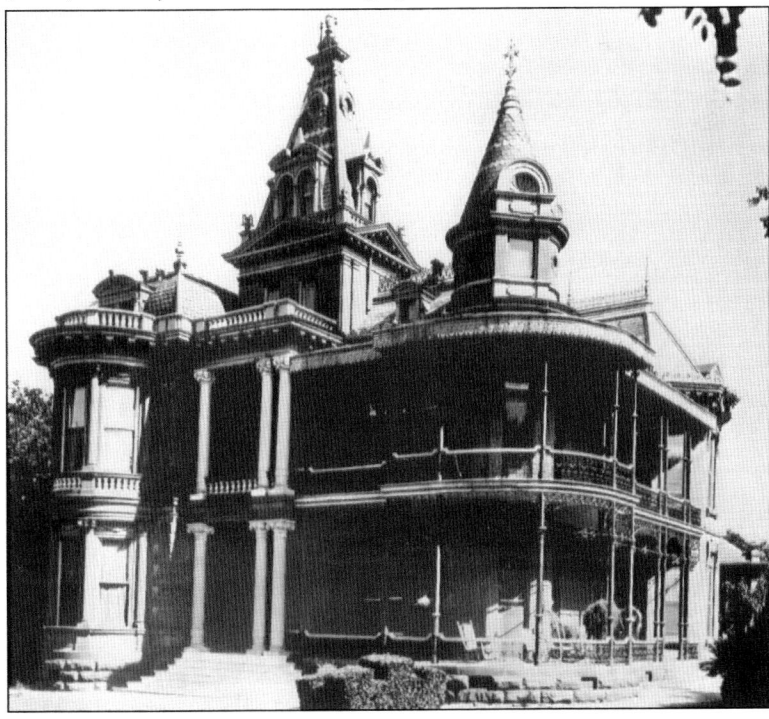

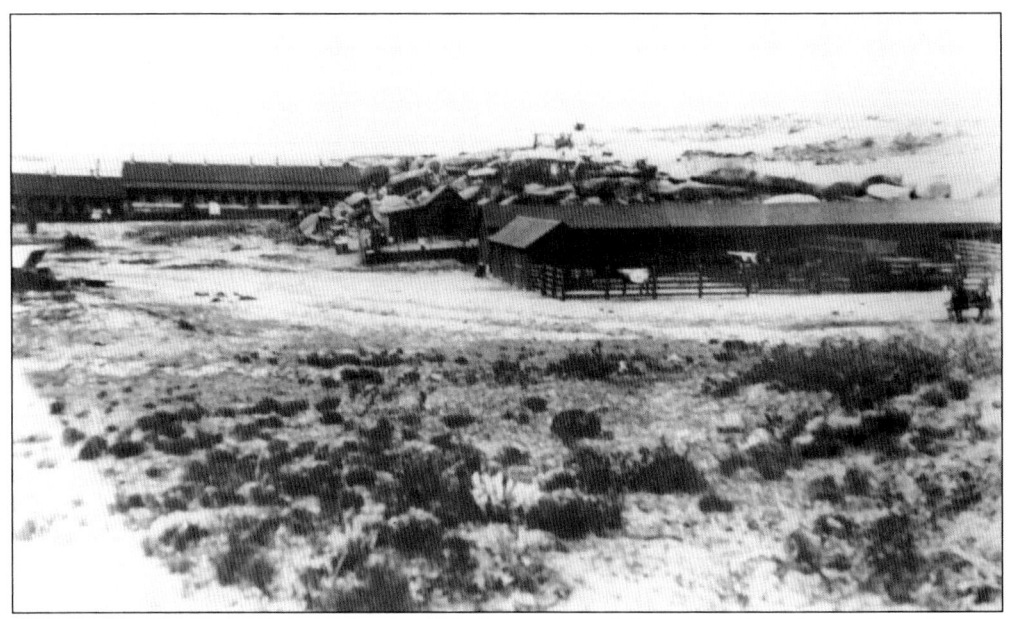

"No matter where you begin on this ranch," observed one LS cowboy, "it's a long ride before you start coming back." W.M.D. Lee and Lucien Scott organized the Lee-Scott Cattle Company, soon grazing 50,000 head of cattle over rangeland nearly the size of Connecticut. Ranch headquarters were twice located at remote sites, but the third compound, seen here, was built four miles south of Tascosa and became the permanent home of the LS. (Photograph by the author.)

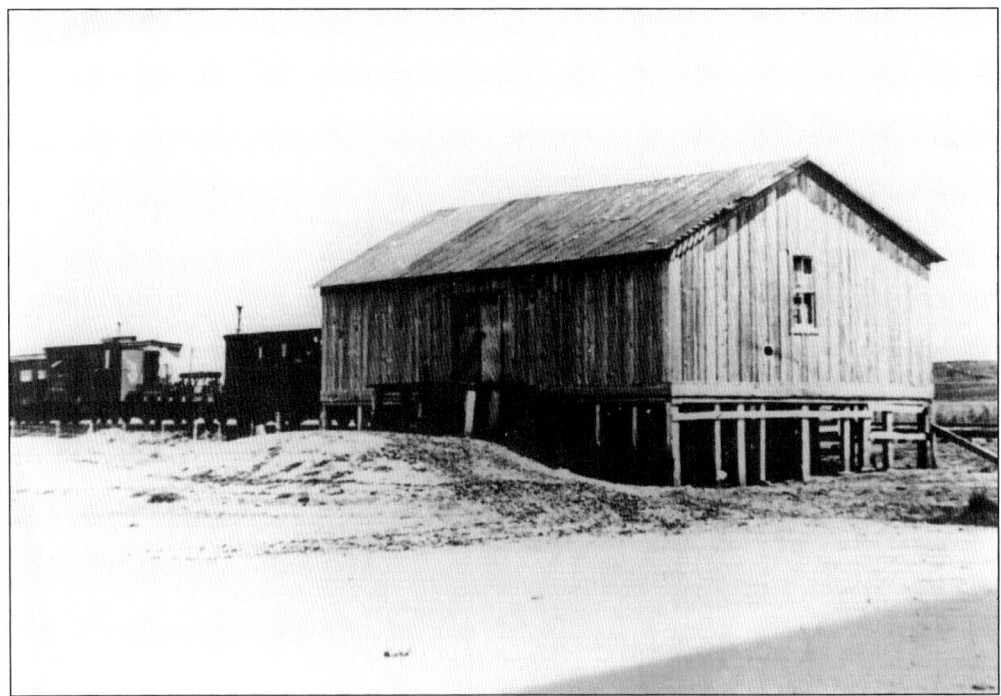

In 1887, the Fort Worth & Denver City Railway passed just south of Tascosa. The LS Ranch, which employed as many as 150 men during its formative years, built a large supply house beside the railroad tracks. (Courtesy Panhandle-Plains Museum, Canyon.)

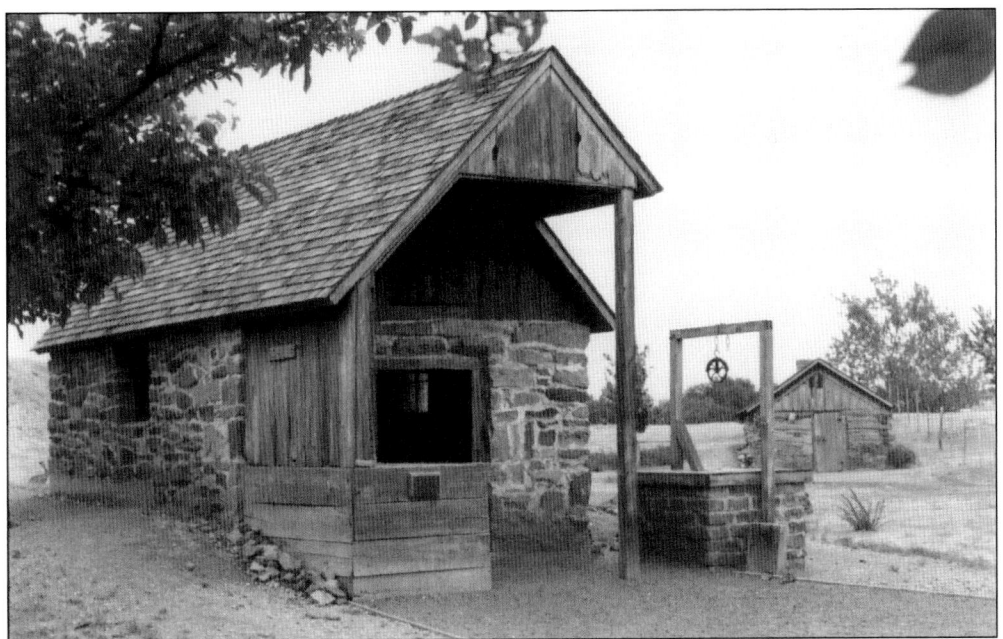

Daniel and Tom Waggoner, father and son, built a pioneer ranch into a million-acre operation with 60,000 head of cattle carrying their famous Three D brand. An early structure was the Waggoner commissary, built in Wichita in 1870, which is now part of the Ranch Heritage Center in Lubbock. (Photograph by the author.)

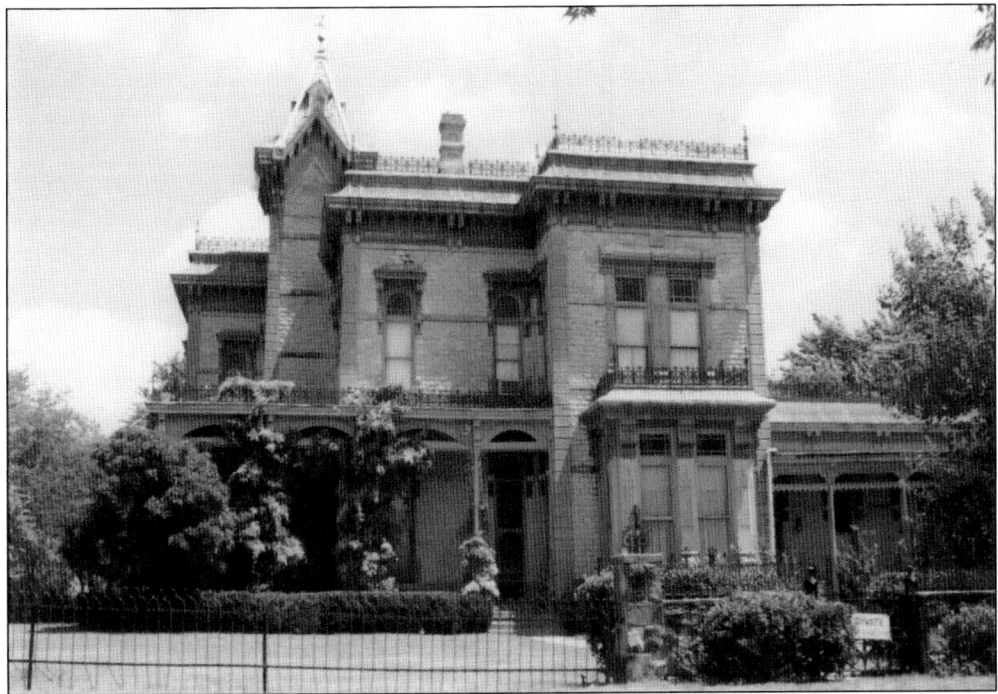

Dan Waggoner built El Castile, seen here, for his growing family. Dominating a hill east of Decatur, the 16-room limestone mansion boasted five marble bathrooms, 16-foot-tall doors, and handsome isinglass fixtures from Denver. (Photograph by the author.)

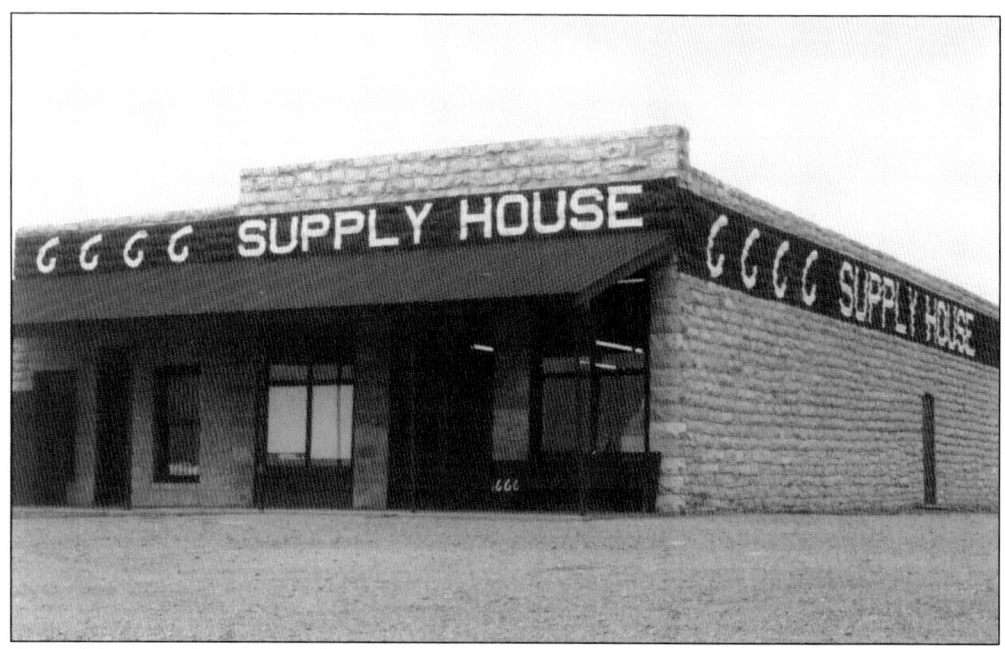

One of the most famous brands in Texas is the "6666" of pioneer cattleman Burke Burnett. In 1868, when he was 19, Burnett rode on a trail drive to Kansas. The next year, he began leading his own trail drives. The oldest building on his Four Sixes Ranch is the commissary, which still stands on the outskirts of tiny Guthrie, just south of the headquarters compound. (Photograph by the author.)

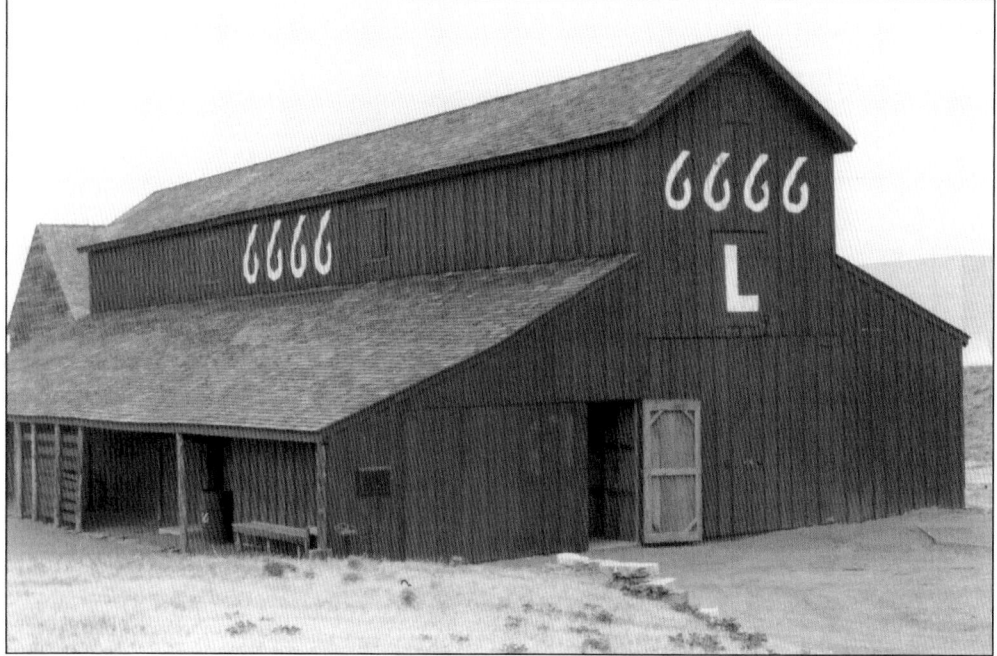

Burke Burnett built this 6666 barn in 1908 at ranch headquarters near Guthrie. The "L" is the 6666 horse brand. The barn has been moved to the Ranch Heritage Center in Lubbock. (Photograph by the author.)

Samuel A. Maverick was an early Texas settler who was elected repeatedly to public office. He engaged in business and land speculation, eventually acquiring over 300,000 acres. Maverick bought a herd of cattle in 1847, but they were allowed to scatter. Unbranded animals in the vicinity were referred to as "Maverick's" or as "a maverick," providing the cattle industry an indelible terminology. (Courtesy Historical and Biographical Record of the Cattle Industry of Texas.)

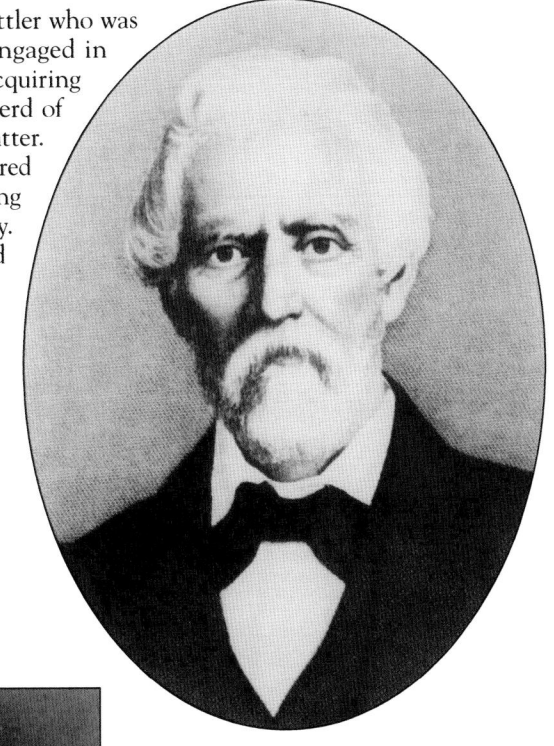

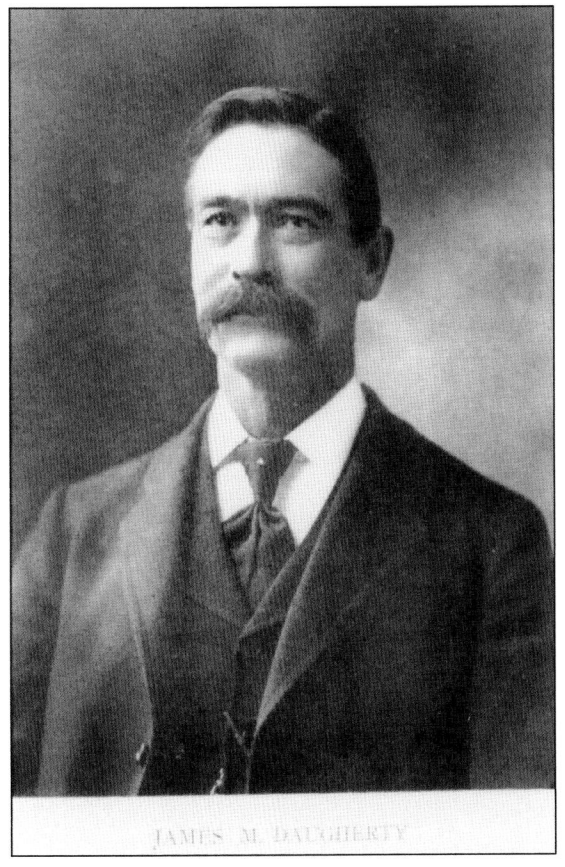

Jim Daugherty became an express rider for Confederate forces at the age of 14 in 1864. In 1866, the adventurous young Texan helped drive one of the first herds of longhorns up the Shawnee Trail. During his career, Daugherty ranched in Texas, Kansas, Colorado, New Mexico, and Indian Territory. By 1900, he controlled 1.5 million acres in West Texas and New Mexico, running more than 50,000 head of cattle. (Courtesy Historical and Biographical Record of the Cattle Industry of Texas.)

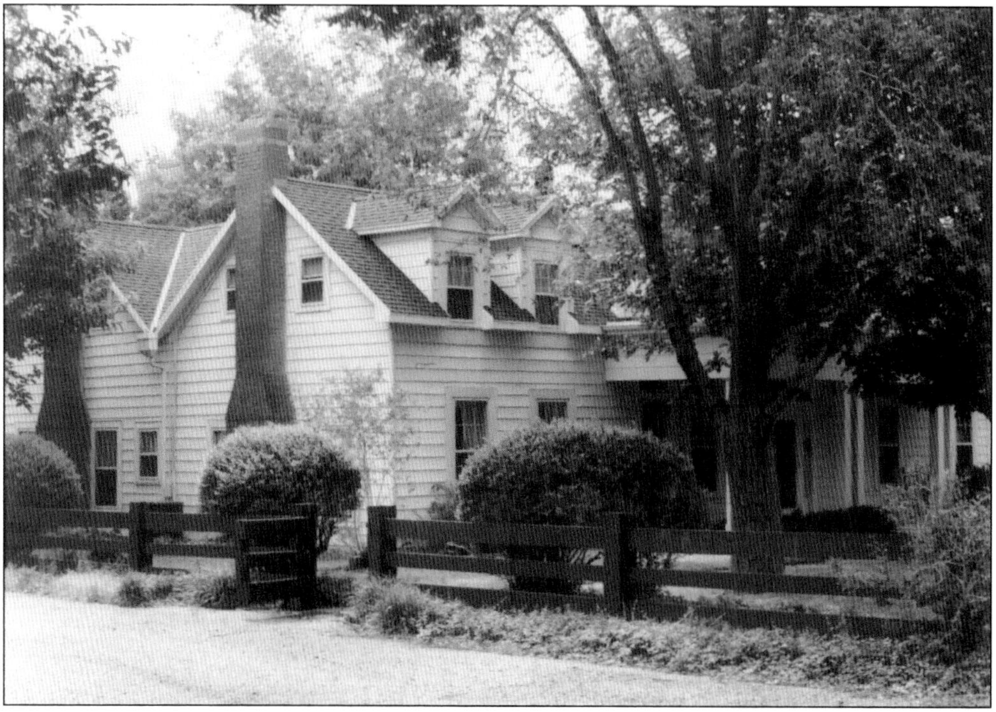

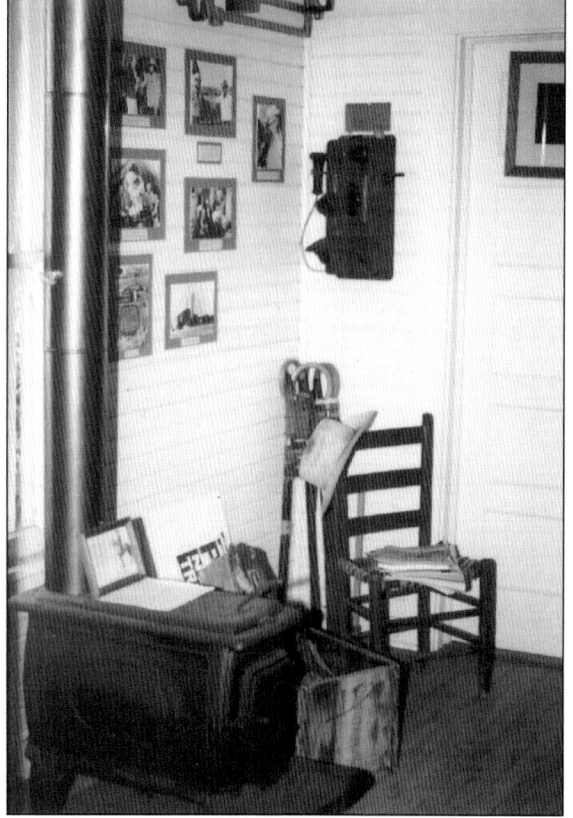

Organized in 1883, the Pitchfork Land and Cattle Company is one of the few ranches from the golden era of the range cattle industry that is still in operation. It is also the only one that is larger today than it was in its frontier heyday. The main residence at the Pitchfork (above) is called the White House. (Photograph by the author.)

The wall phone and wood-burning stove were fixtures of the old Pitchfork office, which has been turned into a museum. (Photograph by the author.)

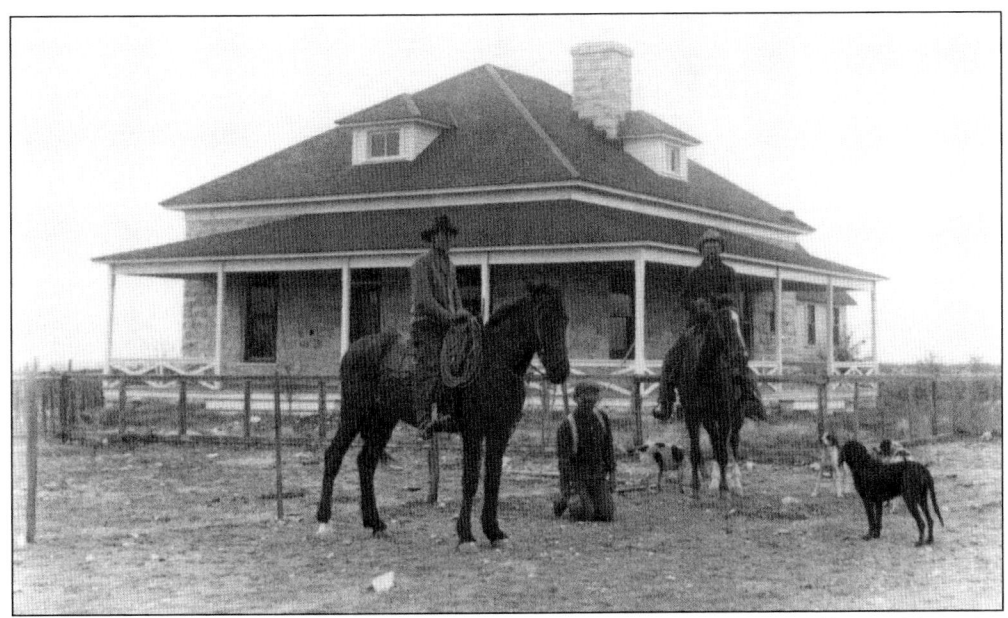

The Renderbrook Spade ranching operation (above and below) was organized by I.L. Elwood, a businessman from DeKalb, Illinois, who manufactured and promoted the popular barbed wire patented by Joseph Glidden in 1874. In 1889, Elwood ventured into West Texas in search of a ranch in which to invest some of his surplus capital. The Renderbrook Spade Ranch, named after the cavalry captain who had discovered a large spring on the property, was purchased, along with another ranch to the south. The first stock herd was bought from J.F. "Spade" Evans, and the brand caused the southern ranch to be dubbed "the Spade." Other property acquisitions expanded the Renderbrook and Spade to a combined 400,000 acres, with a herd numbering 30,000. (Above, author's collection; below, photograph by the author.)

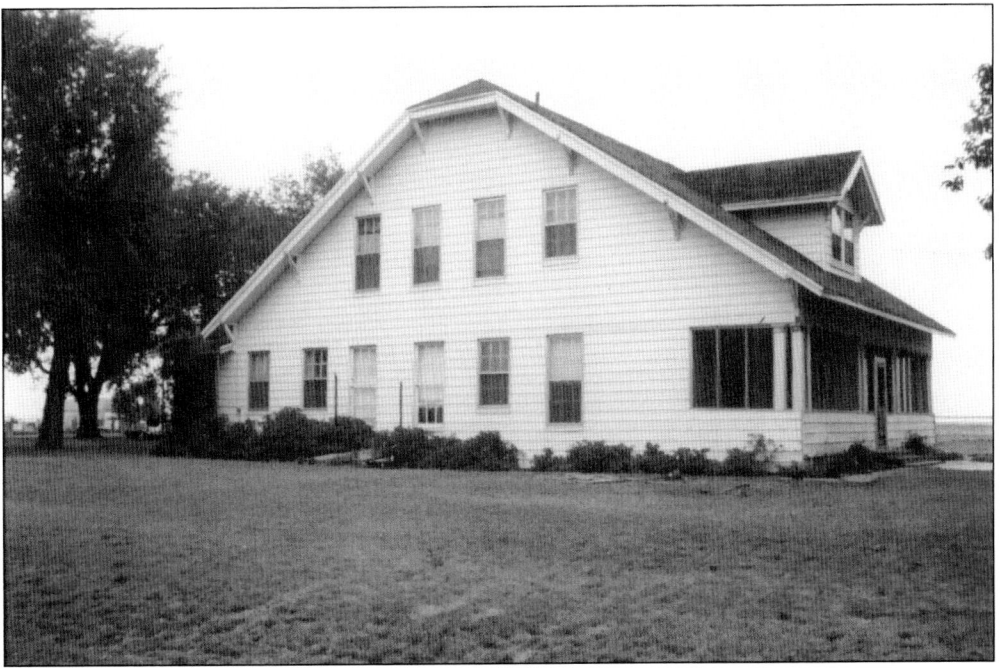

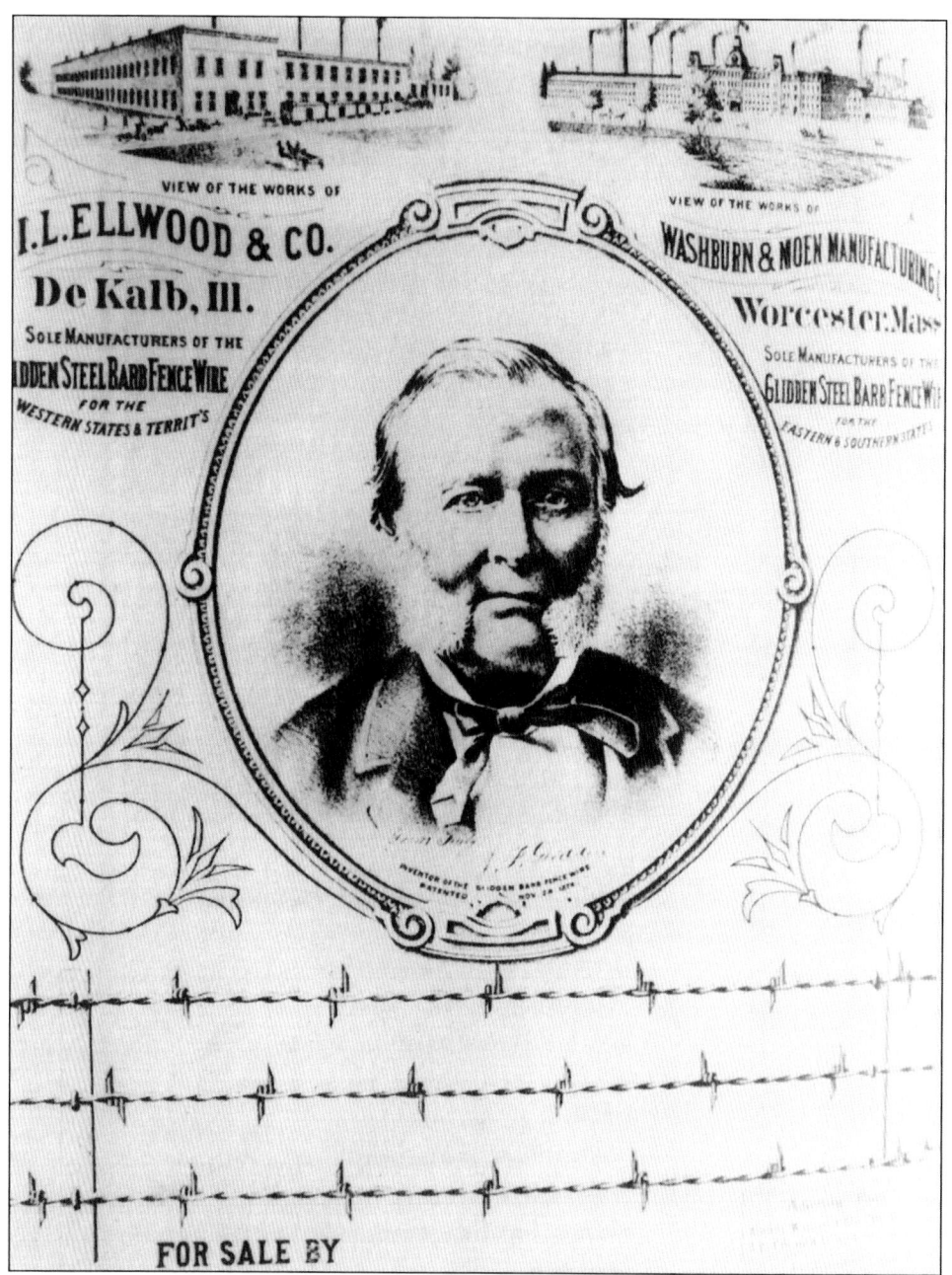

Barbed wire revolutionized the range cattle industry. Longhorns could break through smooth wire, while early barbed wire had barbs large enough to cut plugs from cattle and horses. In the decade after the Civil War, 122 patents were registered, most of which were impractical. But in DeKalb, Illinois, farmer Joseph Glidden devised a simple, inexpensive barbed wire. Merchant I.L. Elwood bought a half-interest in the patent for $265 and built a factory. Annual sales leaped from three million pounds of fencing in 1876 to 50 million pounds in 1879 to 80 million pounds the next year, with a reduction in price from $20 per 100 pounds in 1870 to less than $4 per 100 pounds two decades later. (Courtesy Historical and Biographical Record of the Cattle Industry of Texas.)

Five

Cowboys—
and Cowgirls

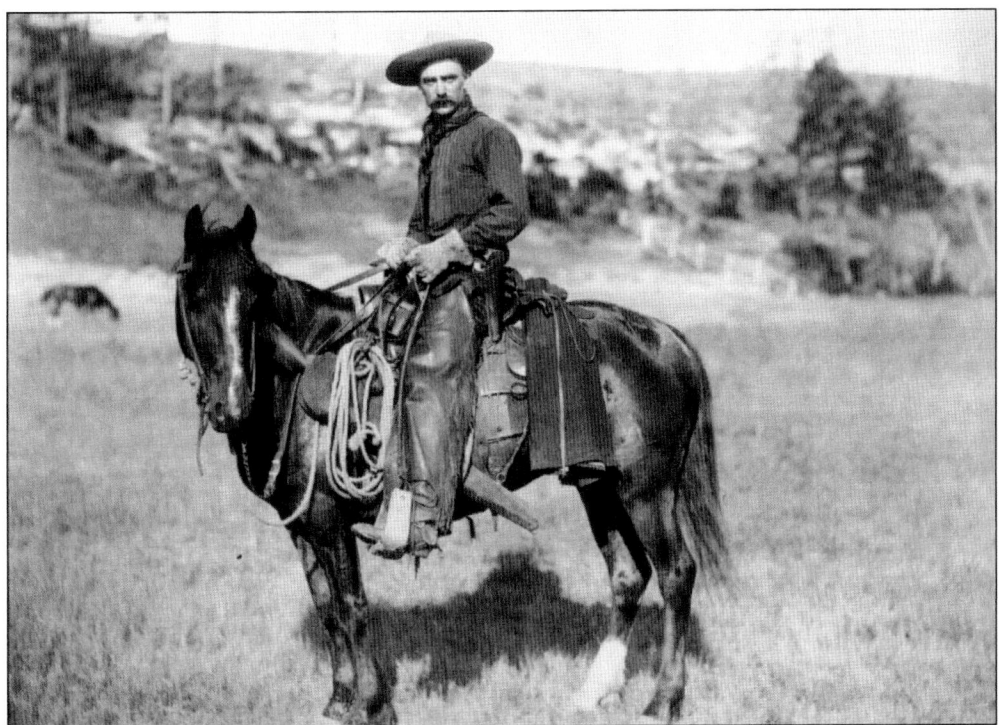

The cowboys of trail drives and open range ranches captivated the popular imagination and became the world's premier folk heroes. This mustachioed cowboy sits astride a western stock saddle in full regalia: high-heeled boots, spurs, fringed leather chaps, riding gauntlets, large bandana, and cowboy hat. A lariat, saddlebags, and a rifle in a scabbard are part of his saddle gear. A revolver is strapped to his left hip, butt forward for a cross draw while in the saddle. He is the image of the frontier cowboy. (Author's collection.)

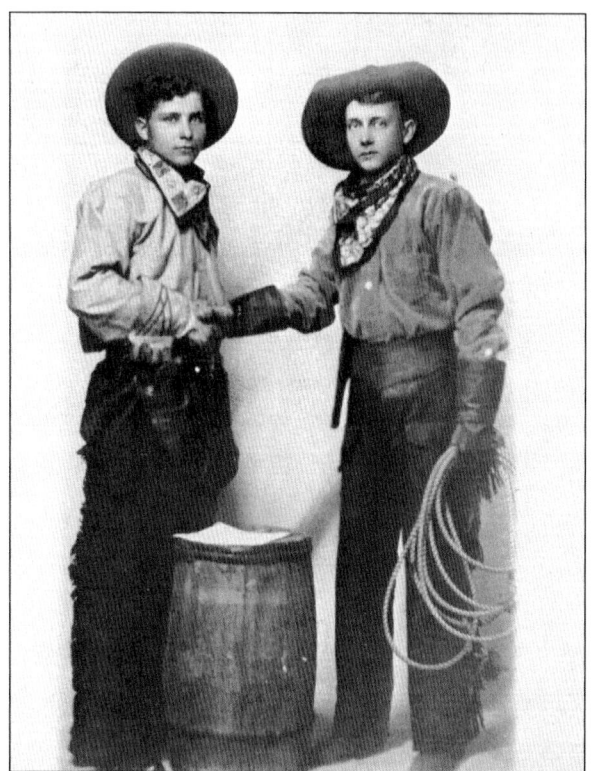

Cowboys loved to pose for photographs in their colorful outfits. These two young men wear chaps, hats, gauntlets, crisp bandanas, a sash, a holstered revolver, a lariat, and even a keg to accentuate the western flavor. Of course, these fellows look a little like pretenders, perhaps easterners who have donned costumes in a photographer's studio. (Author's collection.)

This erstwhile cowboy strikes an intimidating pose with a dangling cigarette and four revolvers—one in each hand, one in his belt, and one in a large holster—but with no bullets in the cartridge belt. (Author's collection.)

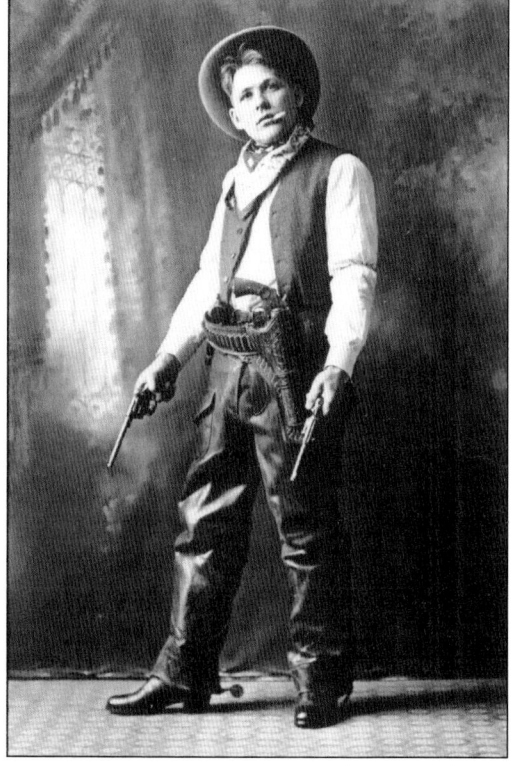

Lampasas County cowboy Will Standard was the son of trail driver Jess Standard. Will began cowboying as a teenager in the 1890s. Like most young cowboys, he liked a good time. Although dressed for the range in boots, chaps, hat, and a holstered six-gun, the tie suggests that he is showing off for the camera and perhaps about to attend a dance in cowboy regalia. (Author's collection.)

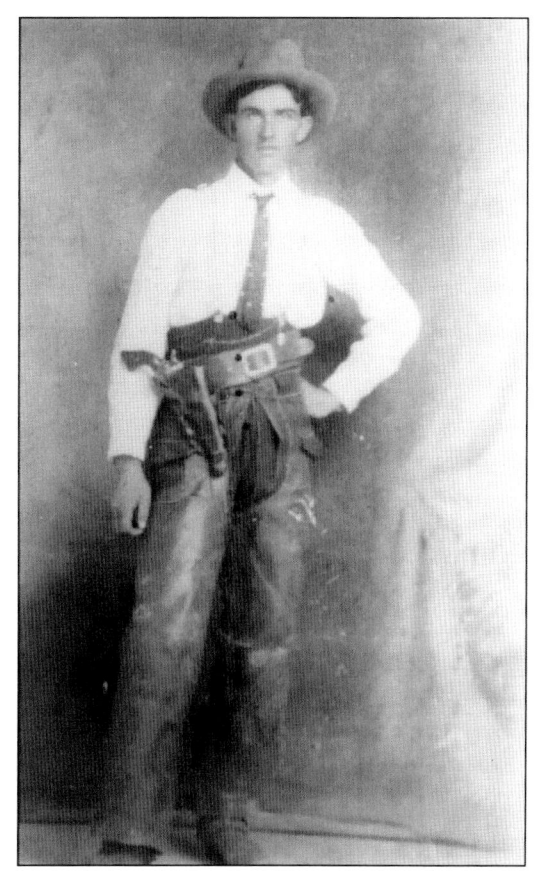

Below, in 1899, these Scurry County cowboys appear to be headed for a party or dance, perhaps in the county seat, Snyder, which became known as a party town, especially around Christmas and New Year's. All seven of these riders have donned suit coats. Some have on vests, several are wearing ties, and at least one man, in the left foreground, shined his boots. (Courtesy Scurry County Museum, Snyder.)

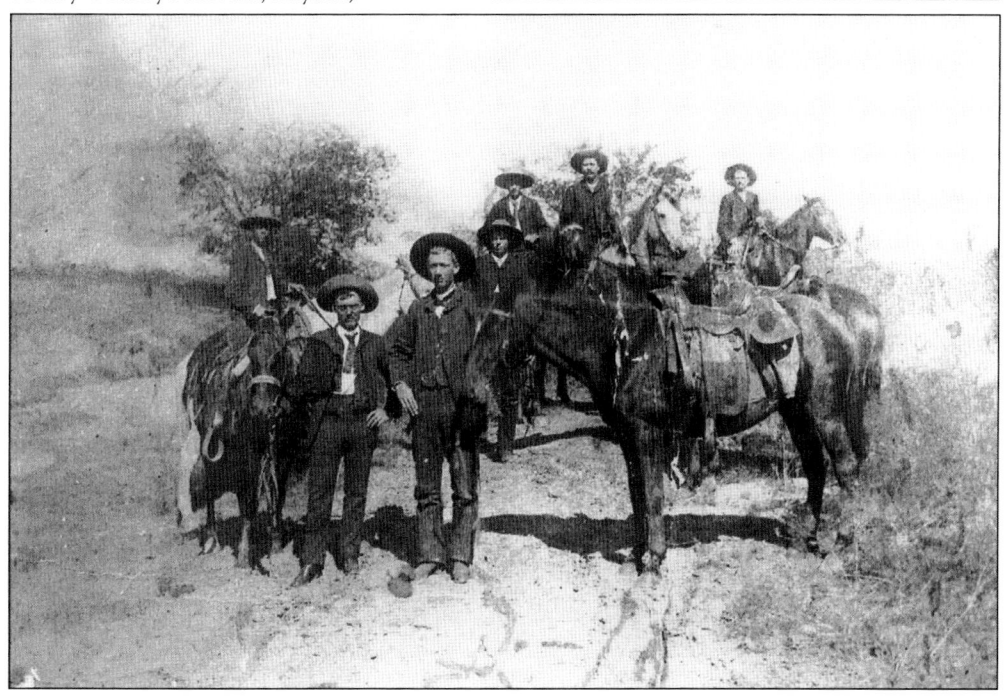

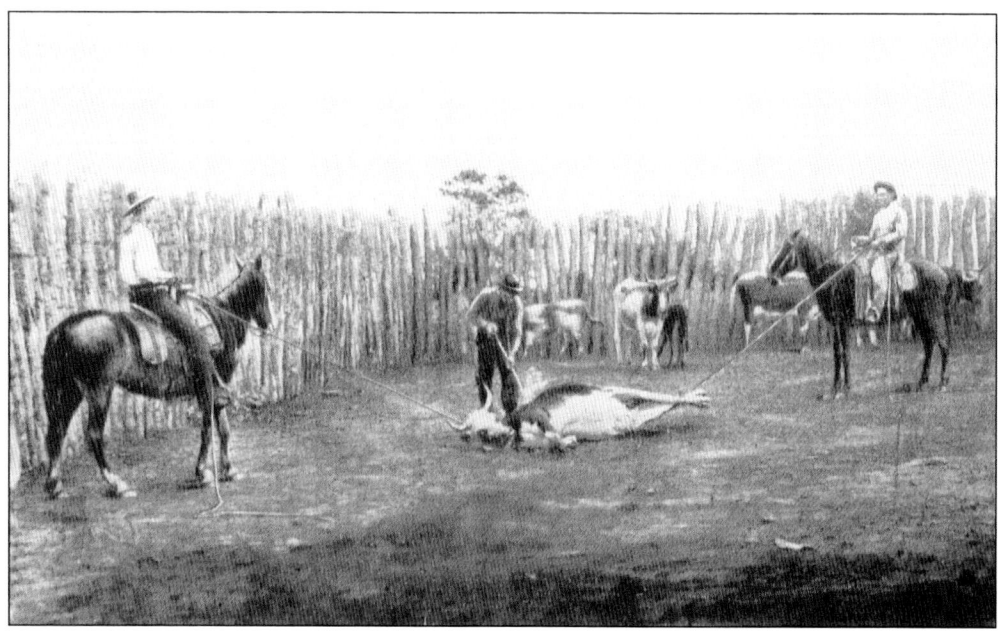

Aside from posing for photographs, cowboys had to work cows, often in a branding pen. In the above scene, a large animal has been "strung out," roped at the head and heels by two cowboys, while a third man applies the branding iron. Below, at a branding pen near Sweetwater, a cowboy has roped a calf while other men brand, cut, and earmark him. (Both, courtesy Historical and Biographical Record of the Cattle Industry in Texas.)

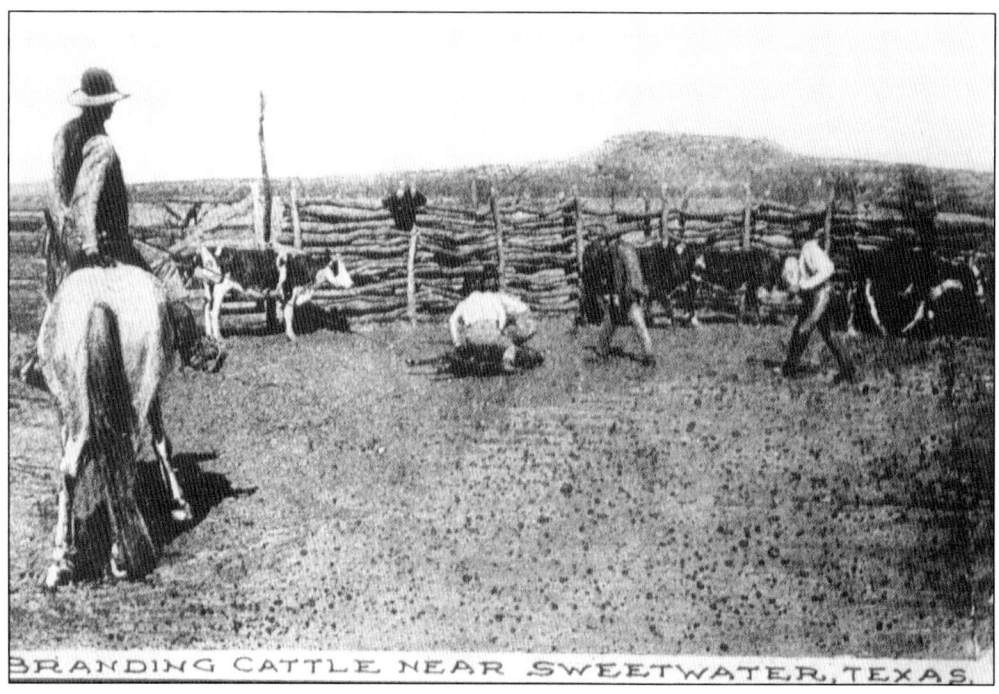

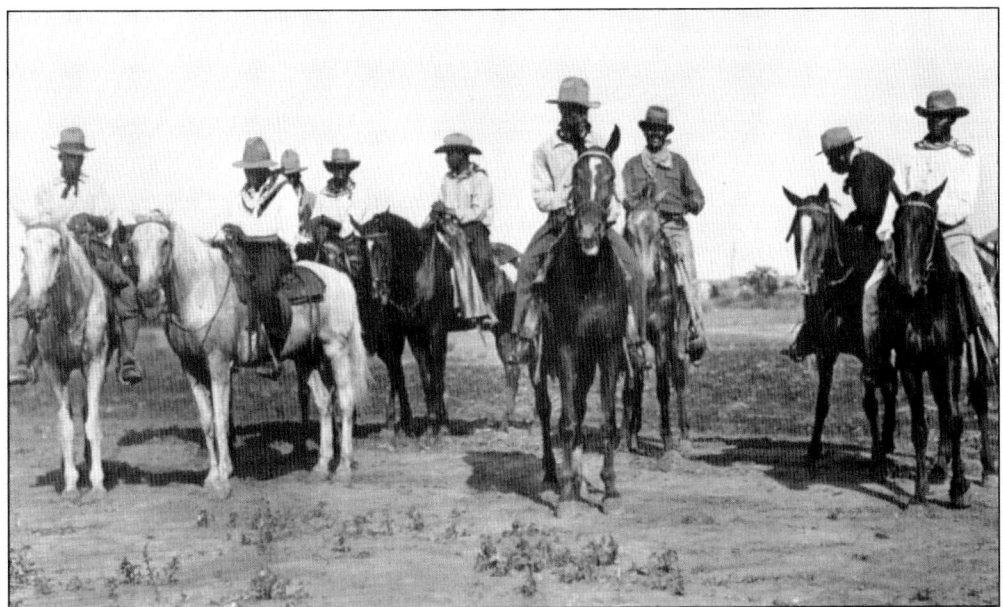

About one in every six or seven cowboys was African American. Most were former slaves, including some who had worked on Texas ranches and learned to ride and handle cattle, skills they used as free cowboys. Black cowboys rarely rose to trail boss or foreman, but they received the same wages as Anglo and Mexican cowboys. (Courtesy Panola College Library, Carthage.)

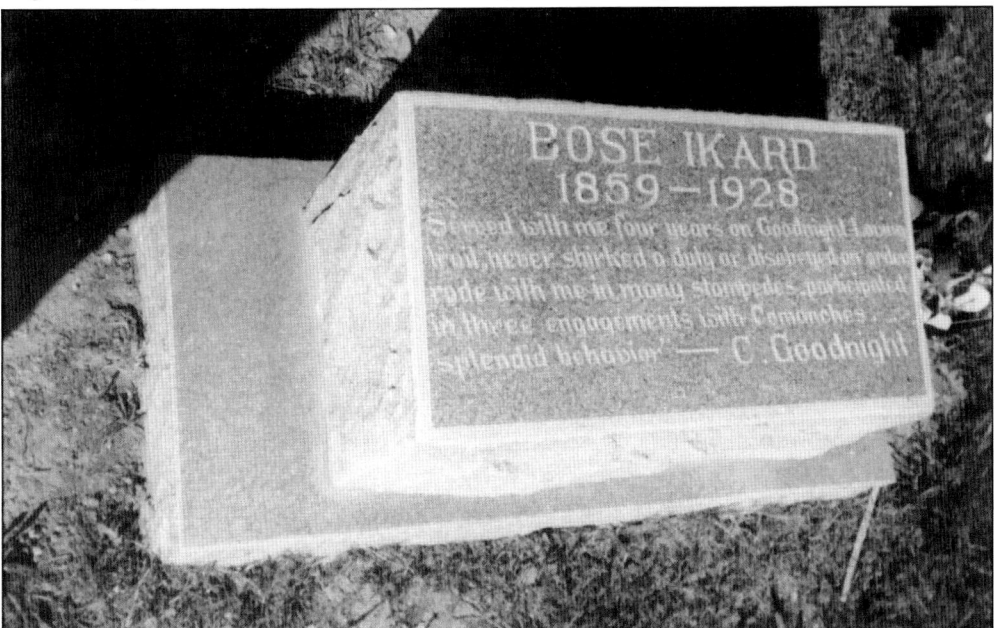

Bose Ikard was a former slave who began working on trail drives with Charles Goodnight in 1866. Although the dates on the tombstone are incorrect, Bose was in his 80s when he died in 1929, and Goodnight provided an inscribed stone for his old companion's Weatherford grave that reads, "Served with me four years on the Goodnight-Loving Trail, never shirked a duty or disobeyed an order, rode with me in many stampedes, participated in three engagements with Comanches, splendid behavior." (Photograph by the author.)

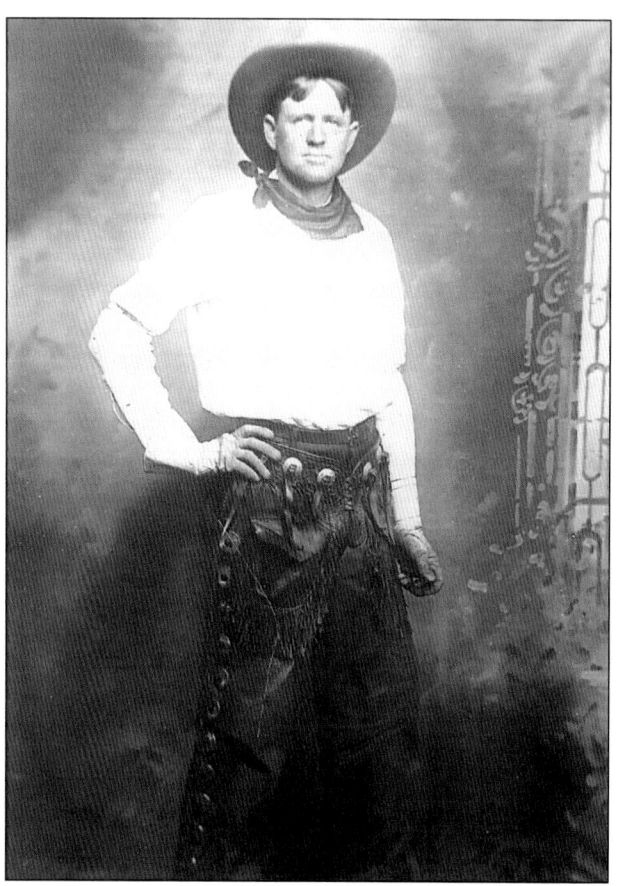

Kelly Sims was one of five sons—there were also five daughters—of a pioneer Kent County rancher. Born in a covered wagon when the family moved to West Texas in 1888, he spent his life ranching the rugged country where he was raised. His outfit featured an especially handsome pair of chaps. (Courtesy Ed Sims, Post, Texas.)

Cowboys nicknamed their spurs "gut hooks," "can openers," "buzz saws," "hell rousers," and "pet makers." Texas cowboys were fond of Lone Star rowels, such as those below at the far left and sixth and seventh from the left. This photograph was taken at the Coryell County Museum in Gatesville. (Photograph by the author.)

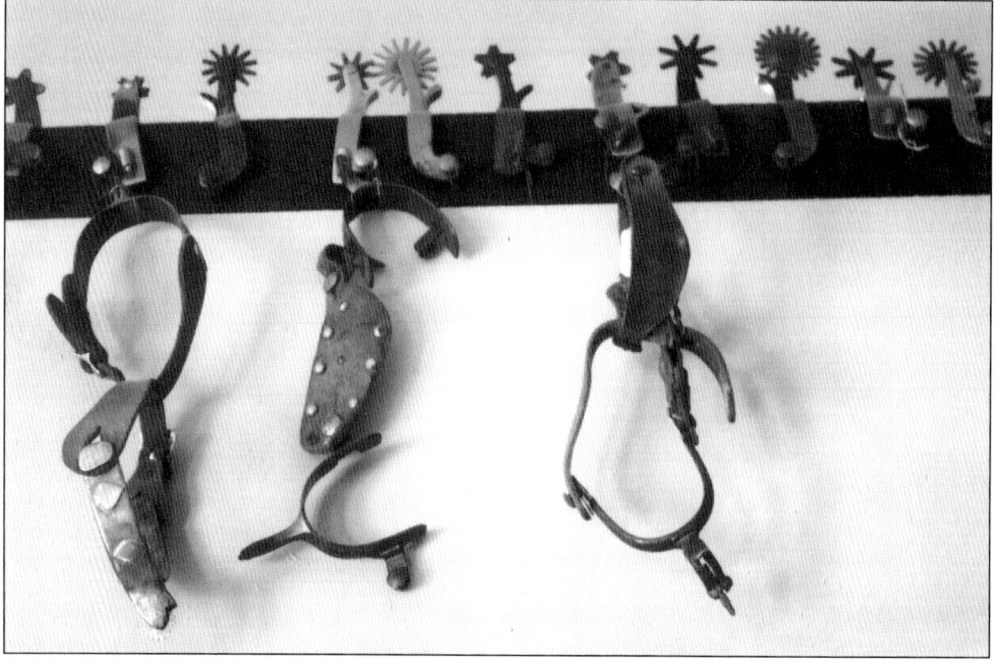

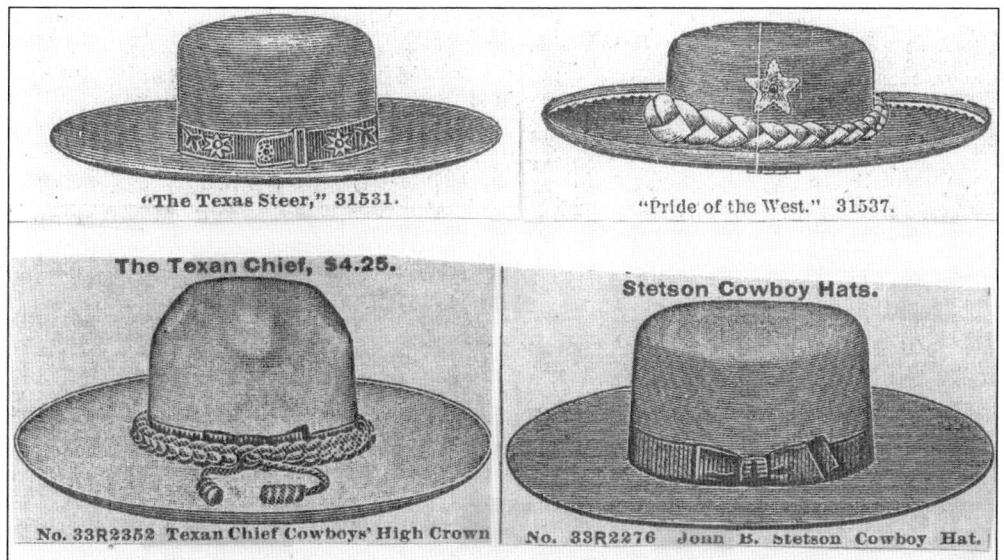

In 1872, Montgomery Ward and Co. of Chicago became the first mail-order firm to sell general merchandise, and in 1886, Sears, Roebuck and Co. began to issue mail-order catalogues. "Monkey Ward" and Sears catalogues were popular on ranches, carrying a wide selection of hats, saddles, ranchmen's vests, firearms, and other items needed in ranch country. (Courtesy Montgomery Ward and Sears Catalogues of, respectively, 1895 and 1902.)

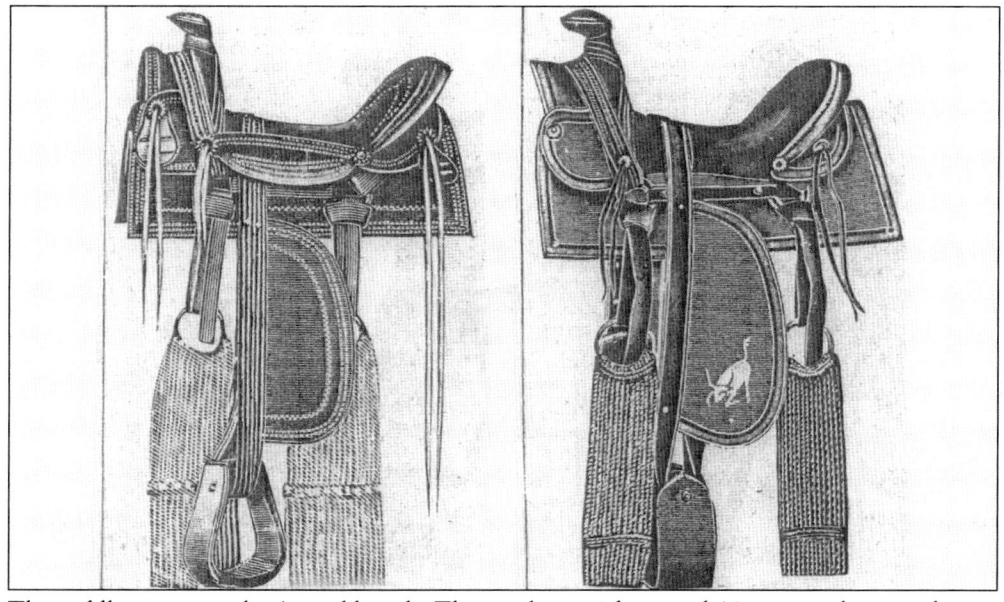

The saddle was a cowboy's workbench. The cowboy might spend 14 or more hours a day on horseback, and light eastern saddles were inadequate for such work. The Montgomery Ward catalogue saddle collection included a "Full Covered Texas Saddle" (left) and a "Double Cinche Rigged Saddle" with "Texas cotton strand cinches" and a longhorn image on the fender. (Courtesy Montgomery Ward Catalogue, 1895.)

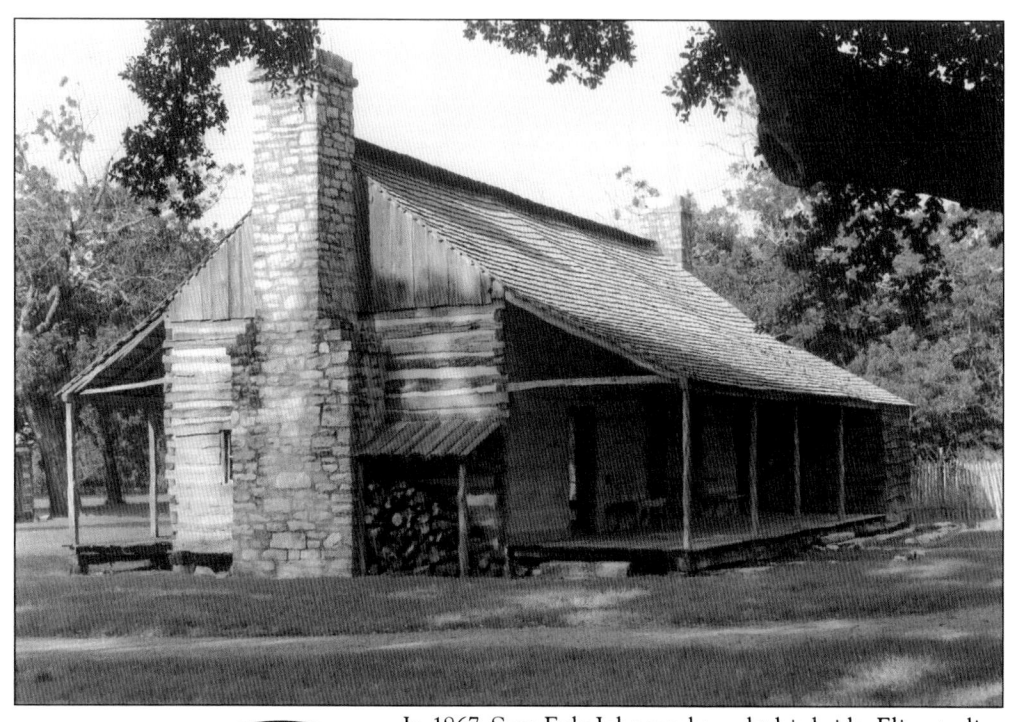

In 1867, Sam Ealy Johnson brought his bride, Eliza, to live in the one-room log cabin above in Blanco County. He added a room and a central shaded passage, creating a classic "dogtrot" cabin. For several years, Sam and his brother Tom bought cattle to drive up the Chisholm Trail to Abilene, Kansas. Both Sam and Eliza would live to see the birth of their grandson, future president Lyndon B. Johnson. (Photograph by the author.)

While Sam Johnson ran his cattle business from horseback, Eliza worked daily to run the home ranch. She cooked meals, hauled water, spun wool, made clothes, and tended animals. Her home was self-sufficient, because the nearest store was in Blanco, a day's ride to the south. (Courtesy LBJ National Park.)

In West Texas, few streams ran year-round, and subterranean water was often hundreds of feet deep. But West Texas was ceaselessly swept by brisk winds, a free source of power that was harnessed mechanically into a modern version of the ancient windmill. The Eclipse was a popular model; this Eclipse windmill pumped water on a Scurry County ranch. (Photograph by the author.)

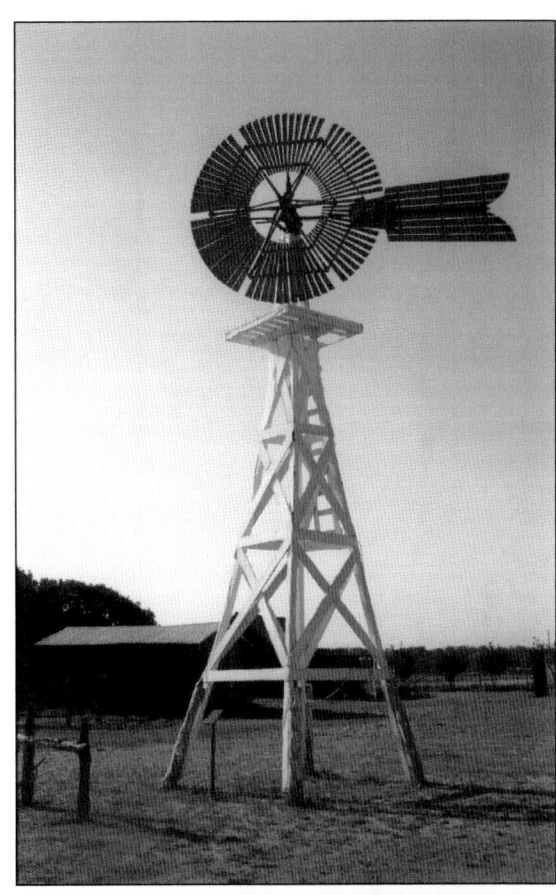

W.H. Ledbetter was an early settler of Shackleford County who dealt with Comanche raids in the 1860s. In the 1870s, he moved his ranching operations northward to the safer vicinity of Fort Griffin. Ledbetter built the dogtrot cabin below with picket construction—with the logs placed vertically instead of horizontally. The cabin is now in Albany. (Photograph by the author.)

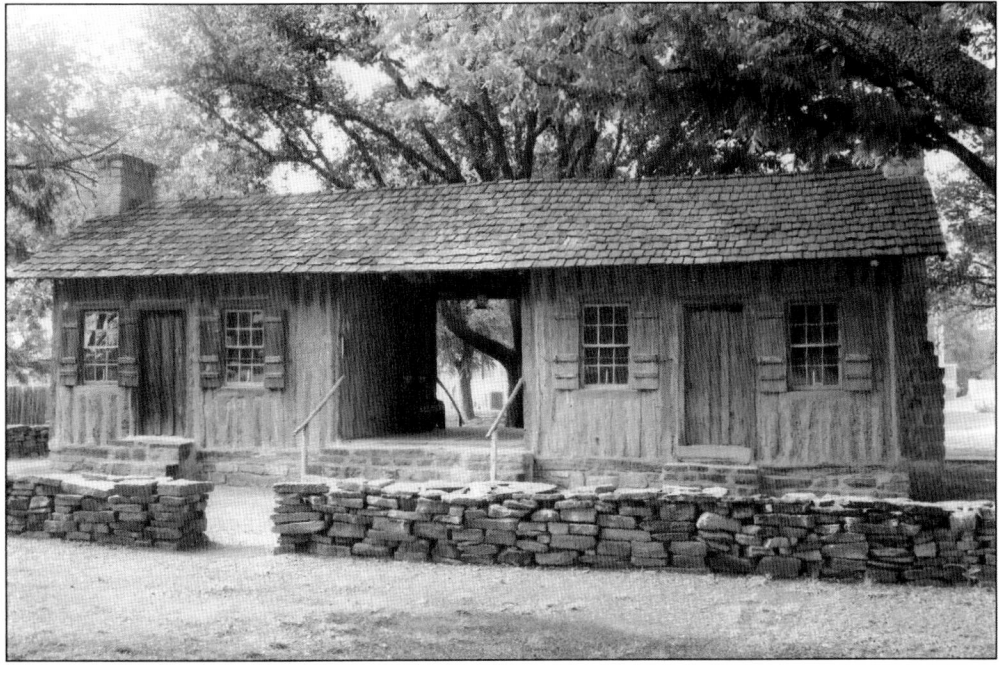

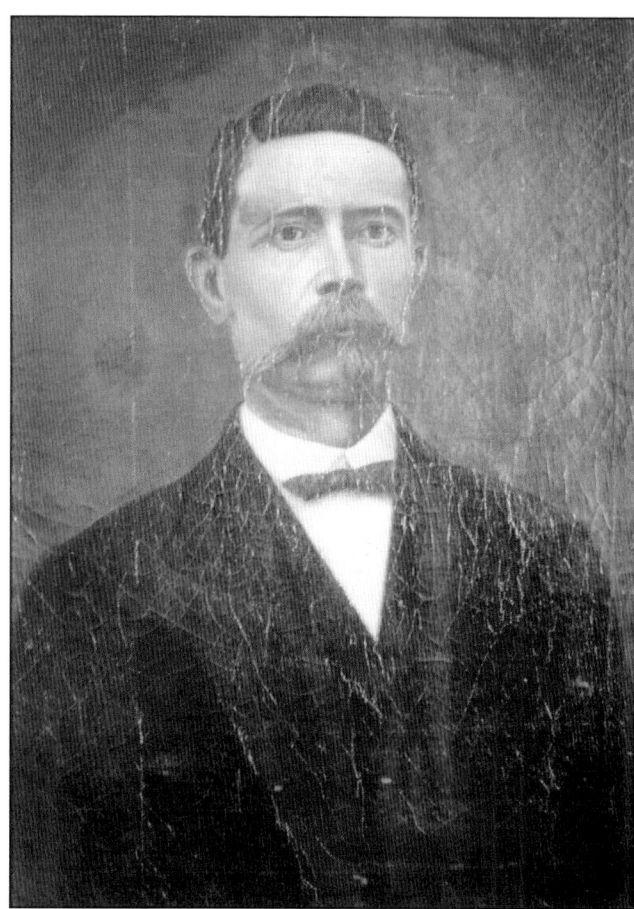

Alex Northington was a Lampasas businessman who became intrigued by the profits possible from selling $4 Texas longhorns for $30 or $40 in distant markets. He invested in a herd of 2,200 longhorns and led a trail drive that ended in Cheyenne, Wyoming. Now committed to the cattleman's lifestyle, Northington traded 400 head of cattle for a 4,000-acre ranch a few miles east of town. He built the ranch house below and moved his family to the spread. Northington remained active in business in Lampasas and was elected to the state legislature, but he lived the rest of his life on his beloved ranch. (Both, courtesy Carol Northington Wright, Lampasas.)

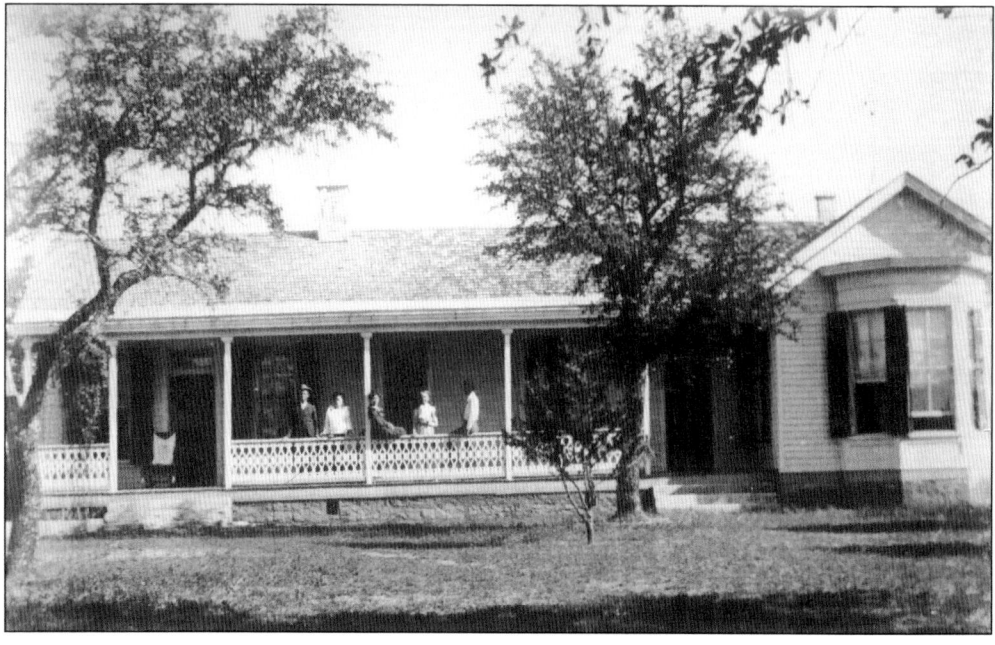

George Jowell's Joly Ranch was located in Palo Pinto County. After Comanches burned his log cabin while he was away on a trail drive, Jowell built this stone structure. The lower room was a kitchen, and at night, he climbed into his bedroom above and pulled up the ladder. This unique ranch house now stands at the Ranch Heritage Center in Lubbock. (Photograph by the author.)

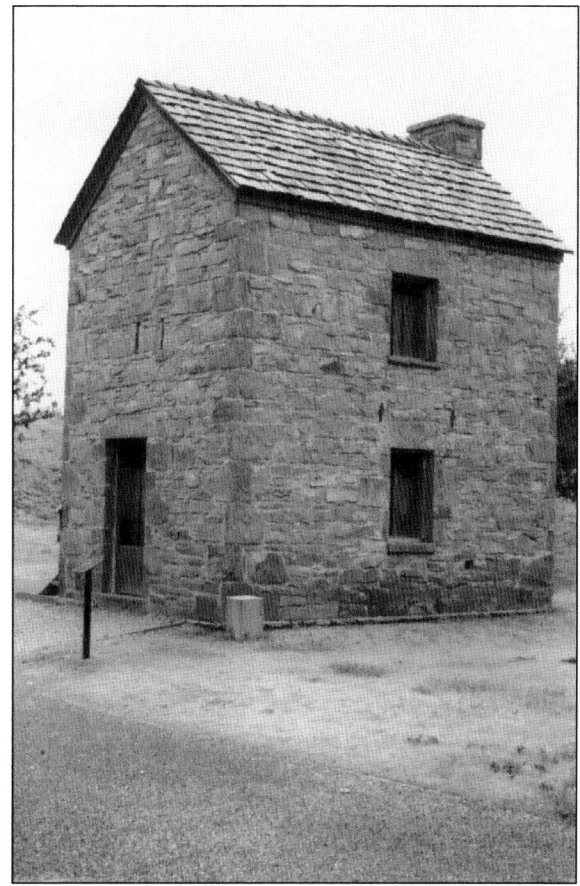

The Fuller Ranch (below) was a Scurry County spread with a substantial headquarters complex built on a hill and a windmill standing high to catch the wind. (Courtesy Scurry County Museum, Snyder.)

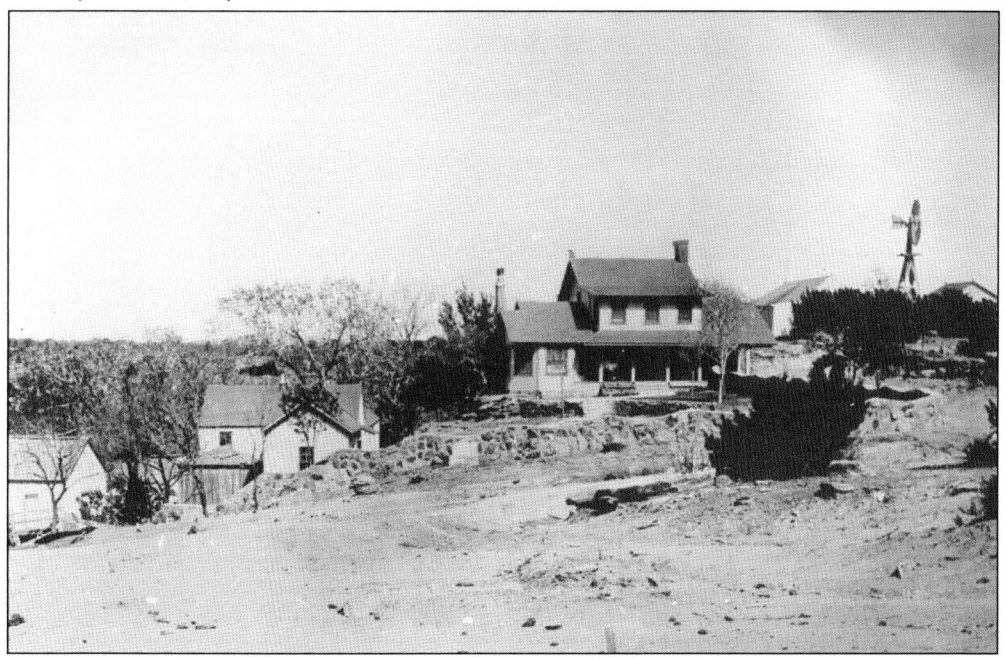

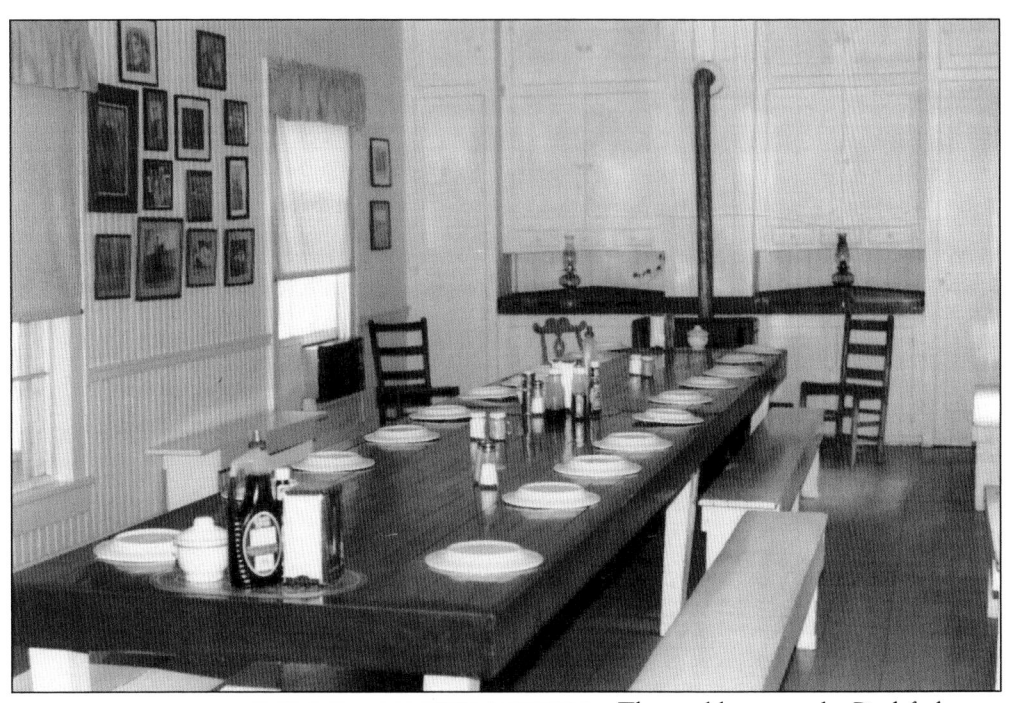

The cookhouse at the Pitchfork Ranch (above) can accommodate 32 hungry cowboys. The dinner bell outside the cookhouse rings at 5:45 a.m. and 11:45 a.m. In the evenings, coyotes come up to eat garbage thrown out the kitchen door. (Photograph by the author.)

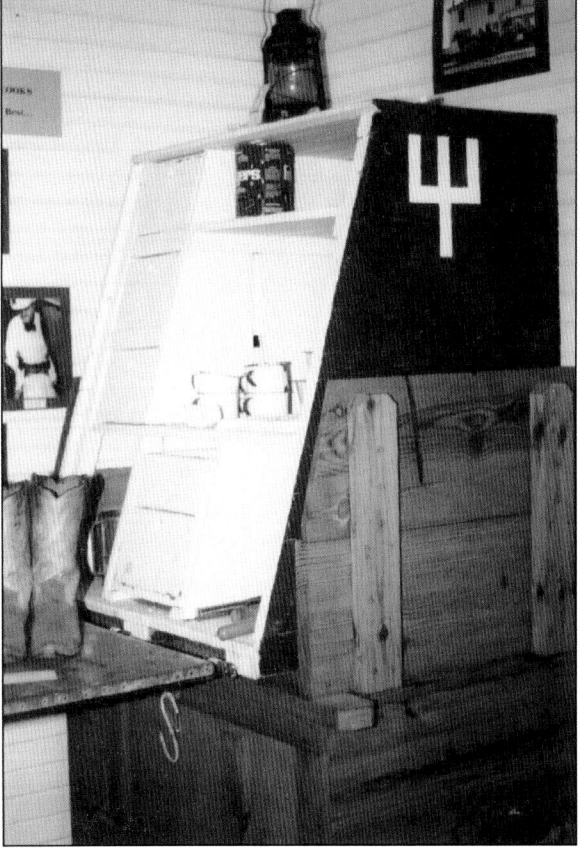

The Pitchfork Ranch still maintains a horse-drawn chuck wagon for roundups in the rugged south range. This old chuck box bearing the distinctive Pitchfork brand is displayed in the office museum. (Photograph by the author.)

"Airtights," or canned food, brought peaches and tomatoes to the chuck wagon and cookhouse. The process of condensing milk was developed by Texan Gail Borden. "Canned cow" was popular with cowboys, and they would say, "No tits to pull, no hay to pitch. Just punch a hole in the son-of-a-bitch." (Courtesy Fort Griffin State Park.)

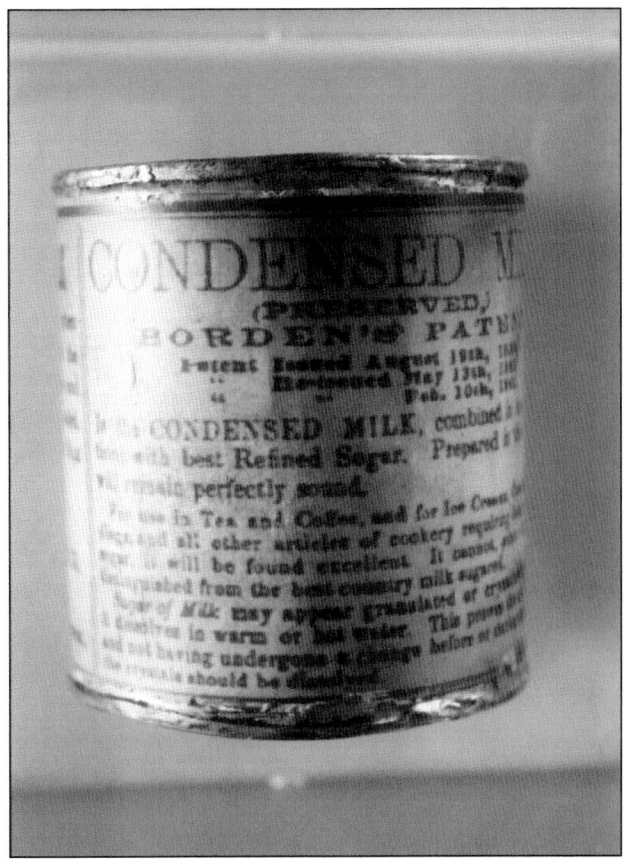

Off-duty cowboys sometimes created intricate rope art. The samples below are on display at the Frontier Times Museum in Bandera. (Photograph by the author.)

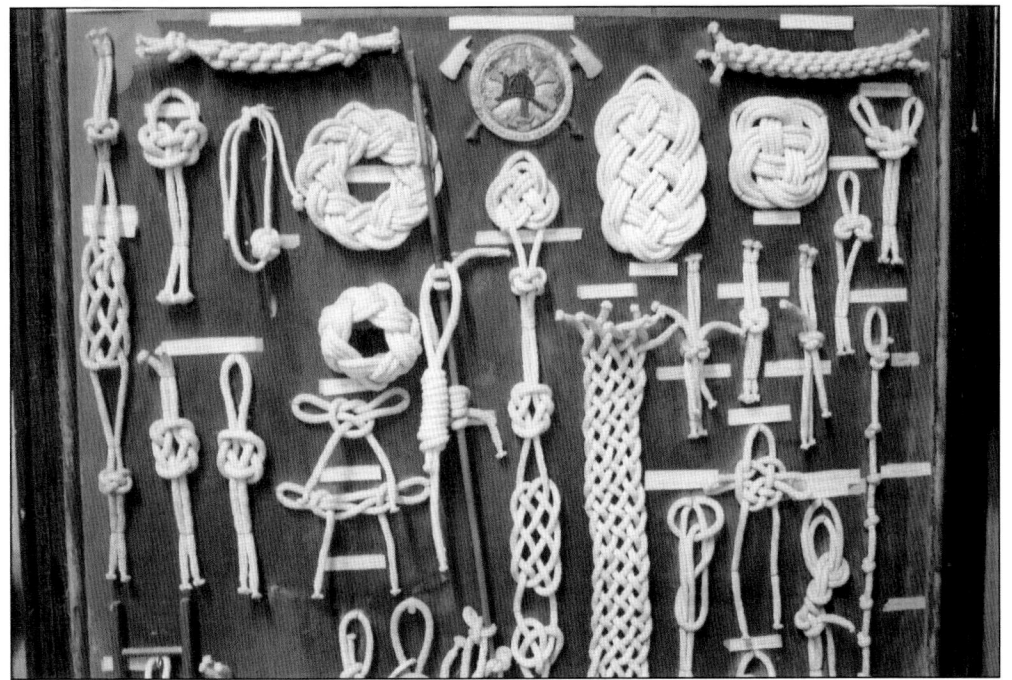

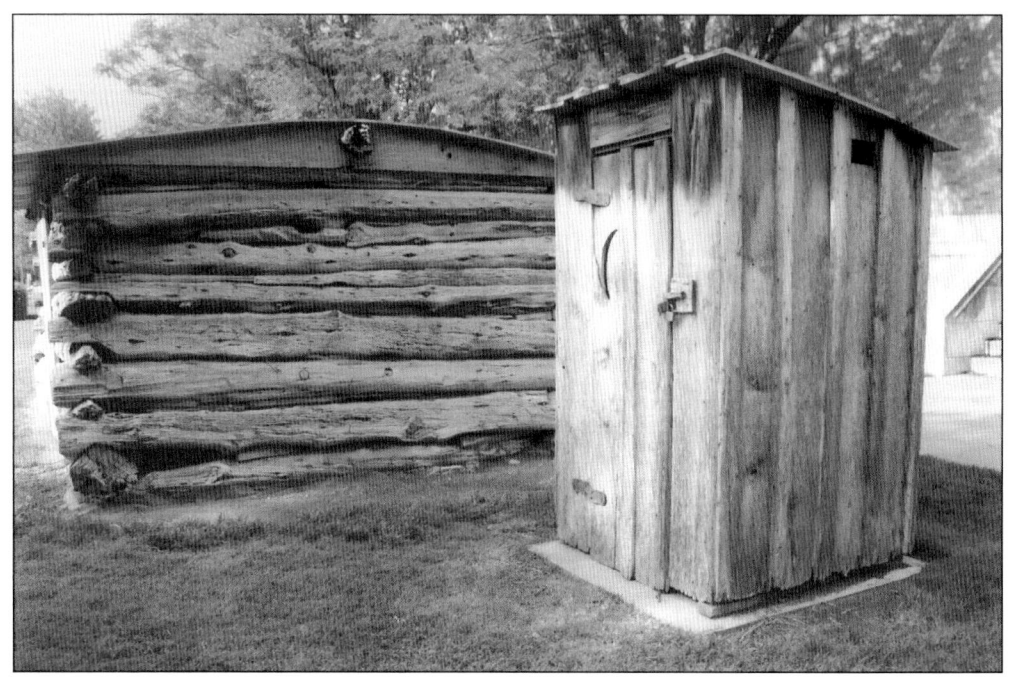

The board-and-batten outhouse above stands behind the T-Anchor Ranch cabin at the Panhandle-Plains Museum. Note the customary moon cut into the door as ventilation. (Photograph by the author.)

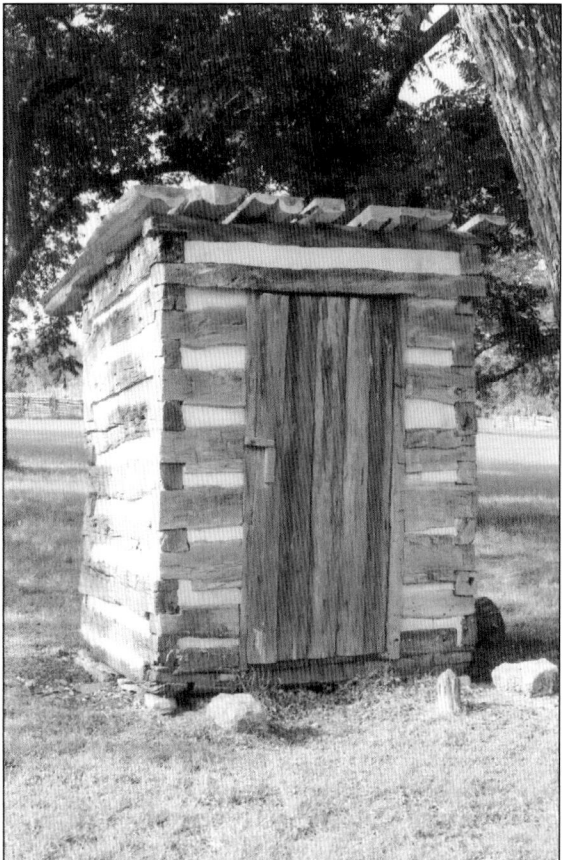

This log outhouse is located near the Johnson ranch cabin at Johnson City. In the absence of "sanitary papers," which were usually unavailable on the frontier, ranch outhouses were equipped with a pile of corncobs or an old "Monkey Ward" catalogue. (Photograph by the author.)

Born in Indiana in 1855, A.B. "Sug" Robertson (upper left) came to Texas as a cowboy. When he was 18, he led eight men and 1,000 head of cattle up the Chisholm Trail. Eventually, he became the manager of a million-acre ranch in West Texas and New Mexico. He and his wife, Emma (upper right), had a daughter and three sons. (Author's collection.)

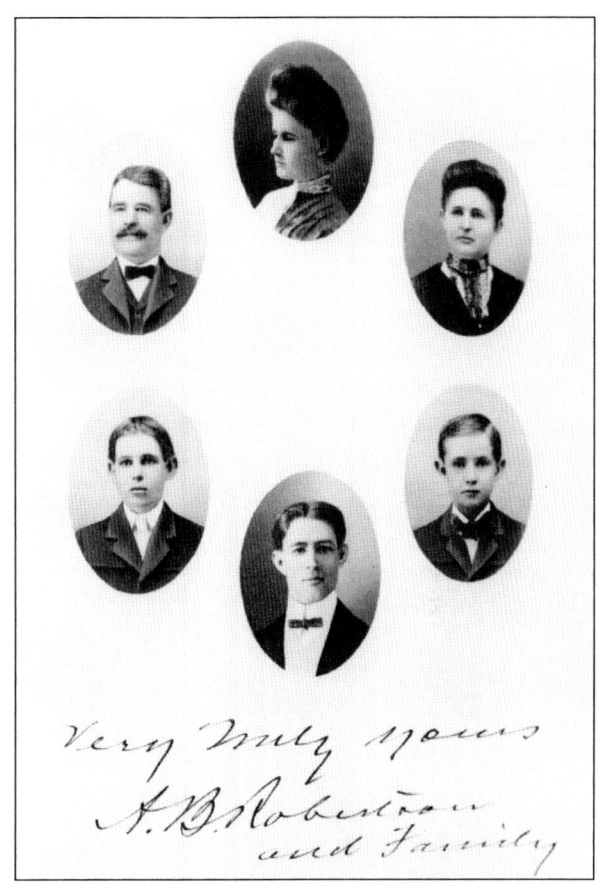

Scurry County rancher Billy Johnson taught all four of his children to ride and shoot. Below, his three sons pose with their horses and two ranch hands in front of the frame ranch house. The oldest son, Emmett, is at left, while young Sidney stands beside his white pony. Joe, the middle brother, is also mounted. The youngest child, Gladys, also became a fine rider and a good shot. (Author's collection.)

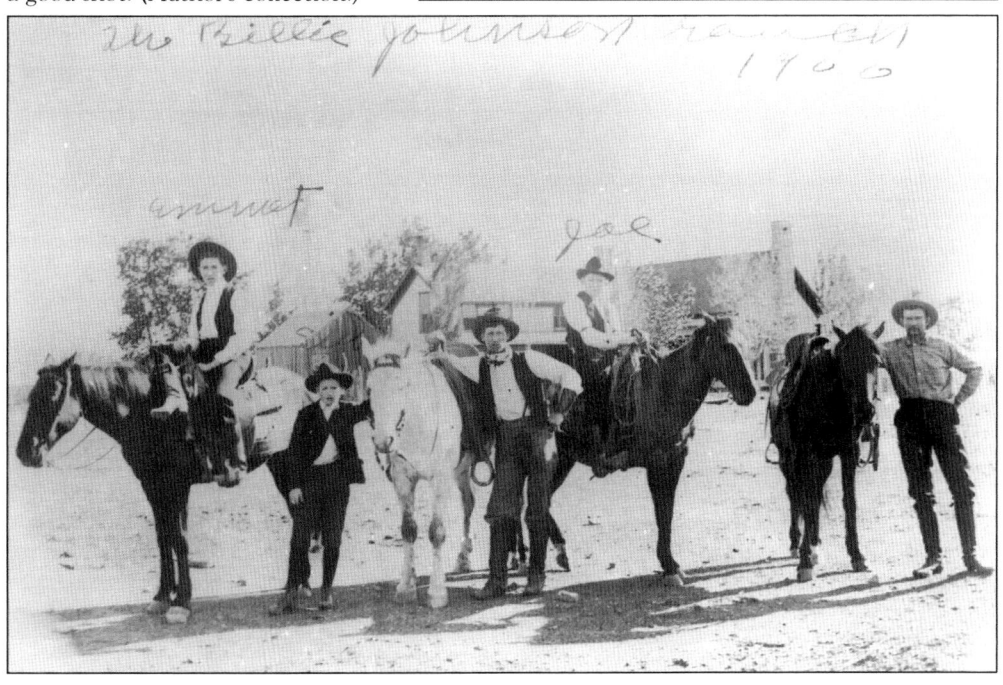

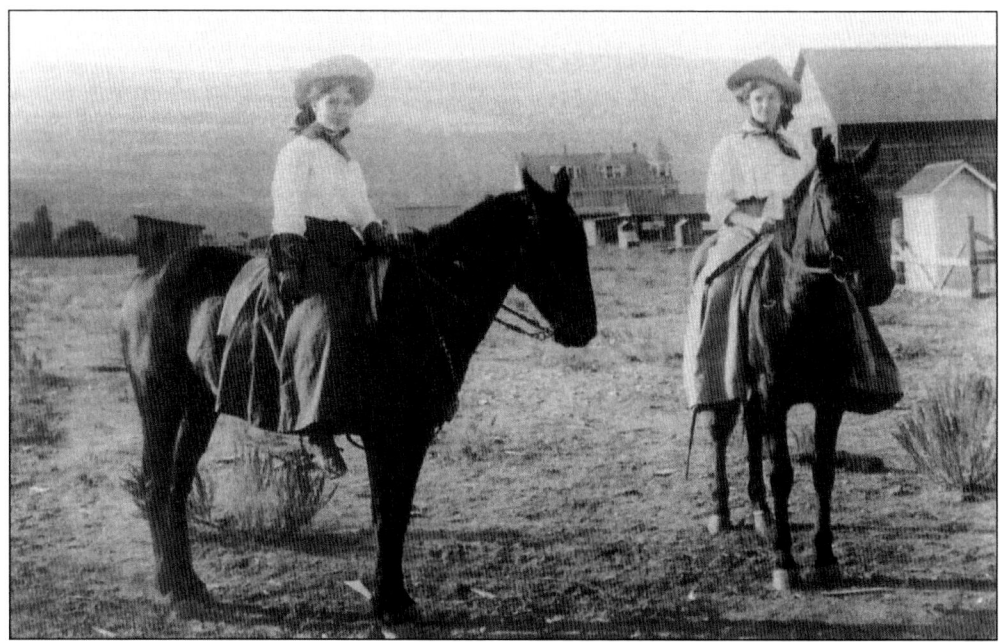

Most cowgirls were rancher's daughters who learned to ride growing up and helped out on the family spread. These cowgirls sport hats, bandanas, and split riding skirts to ride astride. Several cowgirls rode on various trail drives, including Lizzie Johnson Williams, a rancher's daughter who registered her own brand in 1871 and made several drives. (Author's collection.)

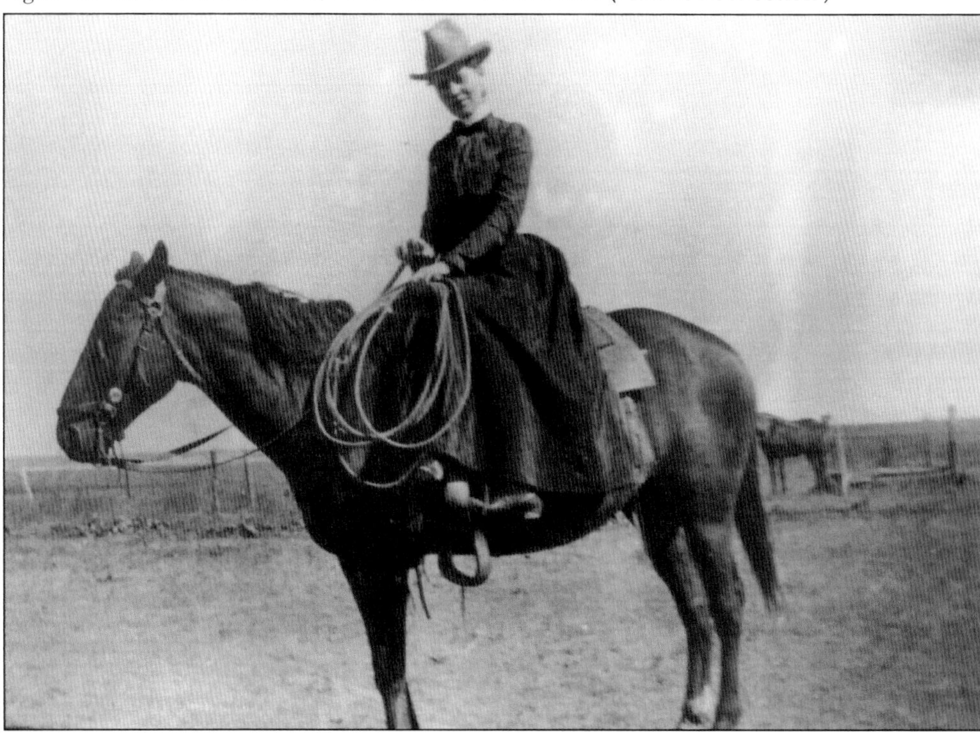

This young ranch woman is mounted sidesaddle, but she wears a tall hat and her lariat is at the ready. (Author's collection.)

Many ranch daughters were taught to shoot as well as ride. This cowgirl wears a revolver and cartridge belt as well as a proper range hat. (Author's collection.)

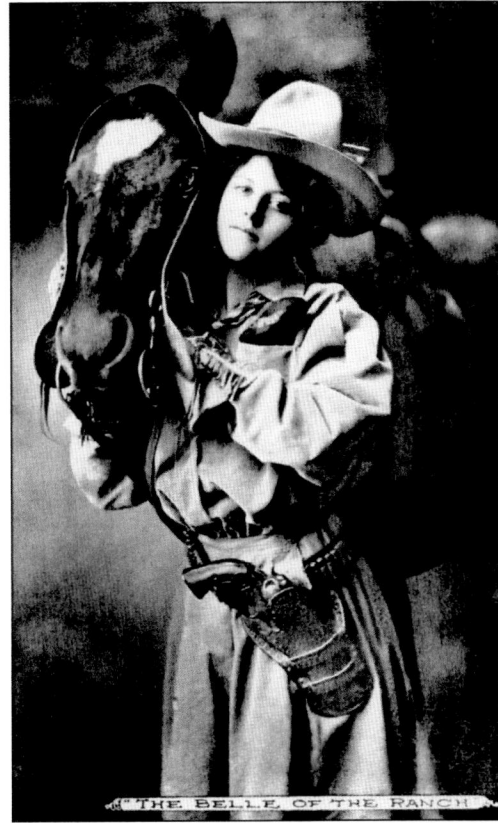

The Scurry County cowgirl below sits erect in the saddle, clad in riding boots, gauntlets, and a big hat. She has masked her face playfully with her bandana, outlaw style, and leveled a small pistol. (Courtesy Scurry County Museum, Snyder.)

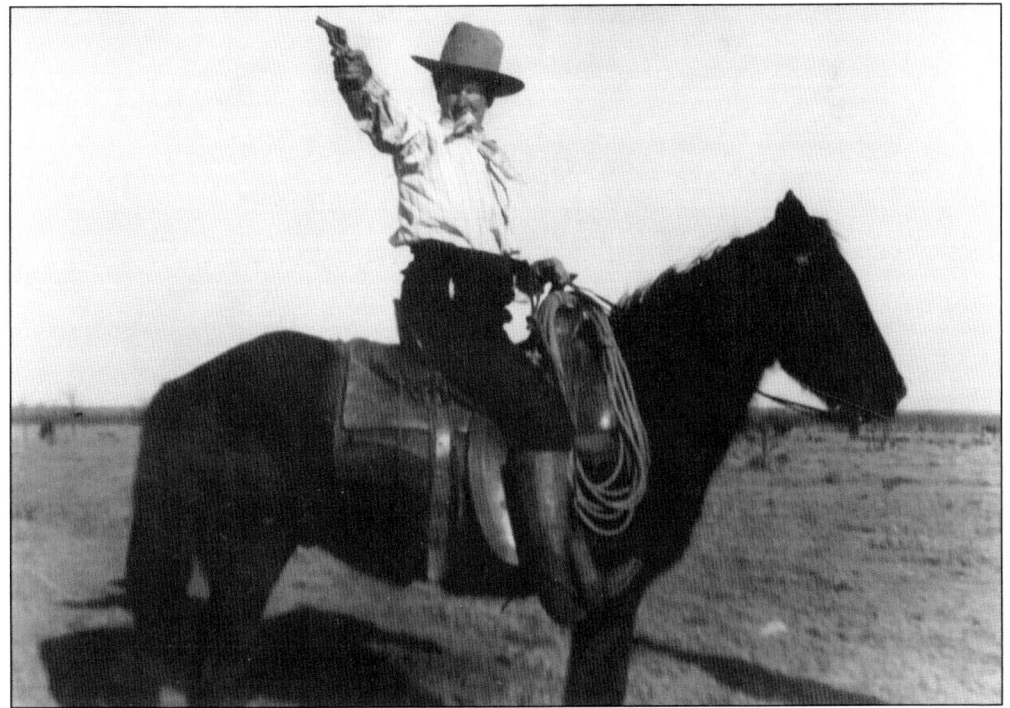

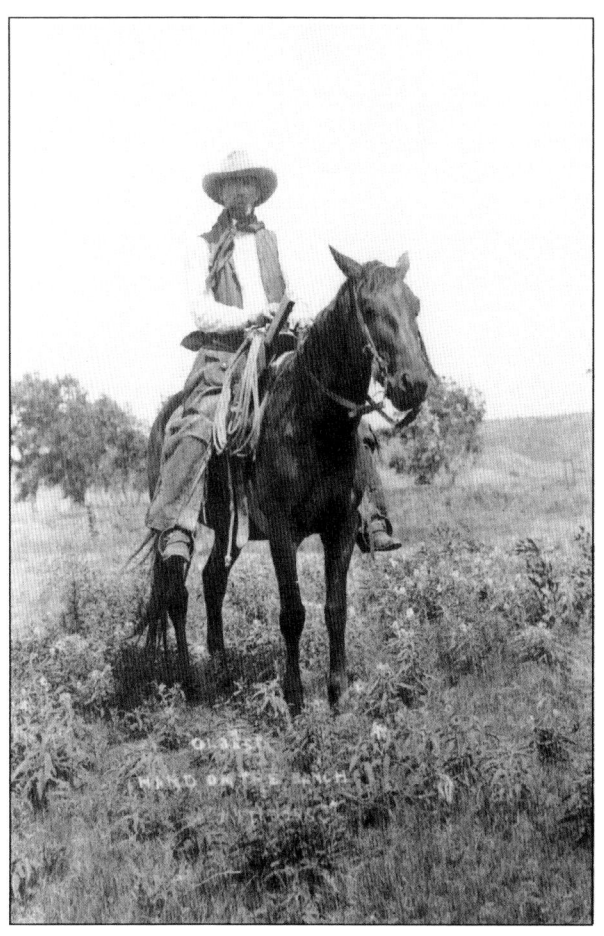

J.W. "Waddy" Peacock was a longtime Matador cowboy who, by the time this photograph was taken, was the oldest hand on the ranch. (Courtesy Panhandle-Plains Historical Museum, Canyon.)

Below, cowboys brand calves on a Scurry County ranch. The calf is pinned down by the knees of two veteran hands. (Courtesy Scurry County Museum, Snyder.)

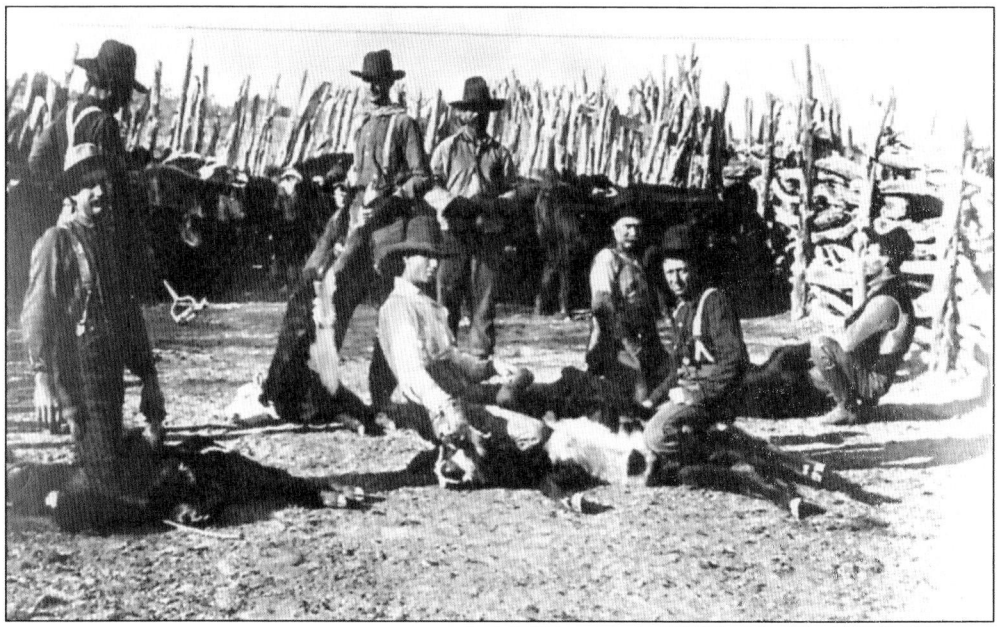

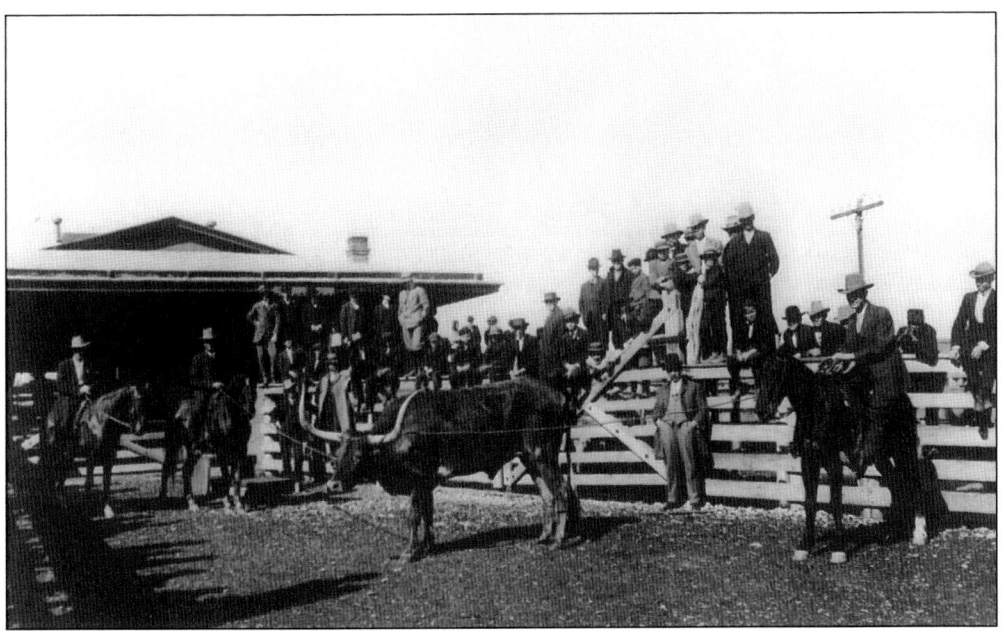

In the early 1900s, this crowd of Scurry County ranchmen gathered to see an old-time longhorn strung out by two horsemen. (Courtesy Scurry County Museum, Snyder.)

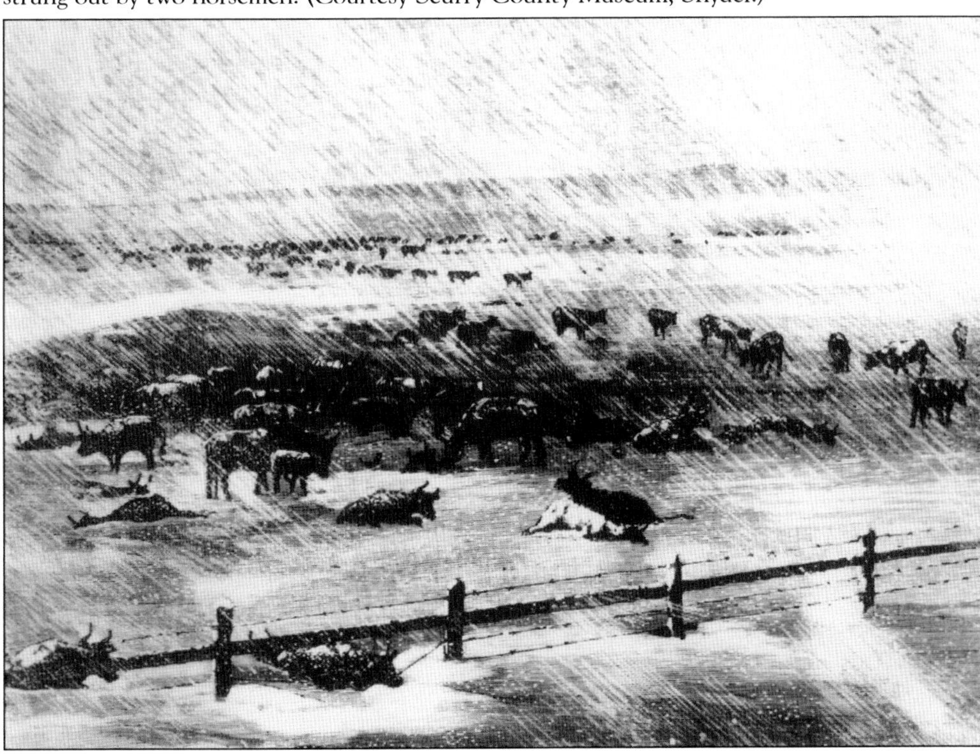

Blizzards dealt harshly with West Texas cattle, especially in the Panhandle. After the open range was fenced in, cattle could no longer drift south during a raging snowstorm; instead, they would pile up against a fence and often die in the snow. (Courtesy Historical and Biographical Record of the Cattle Industry in Texas.)

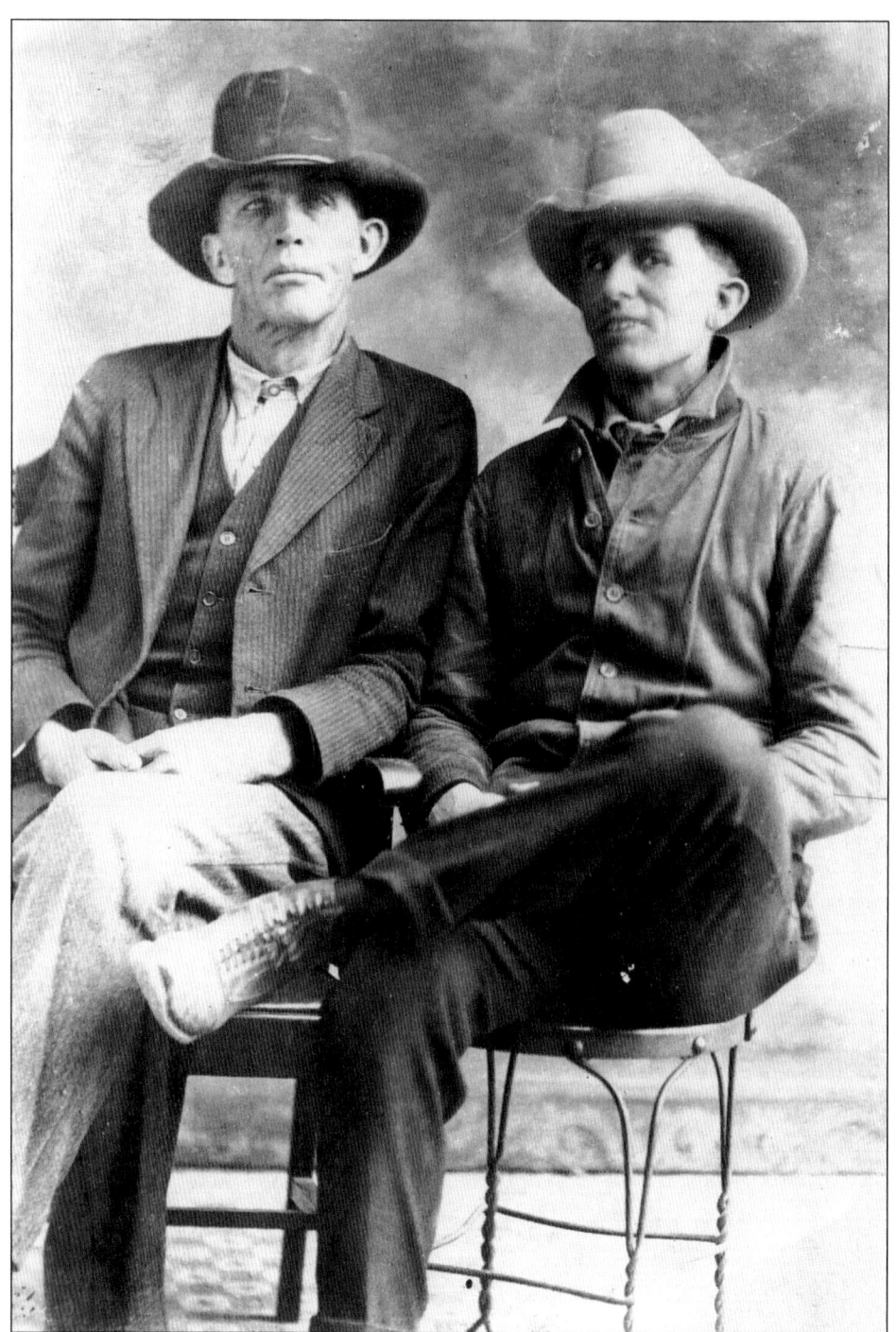
These two old Lampasas County cowboys remained friends even though Will Standard (right) left the range for steady work in town after his marriage. Will and Lucile Standard raised eight children in Lampasas, and one son, Ted, became a rodeo cowboy and master saddle maker. Will trained horses as a sideline, and he eagerly talked about cowboying with old comrades. (Author's collection.)

Six
TEXAS CATTLE TOWNS

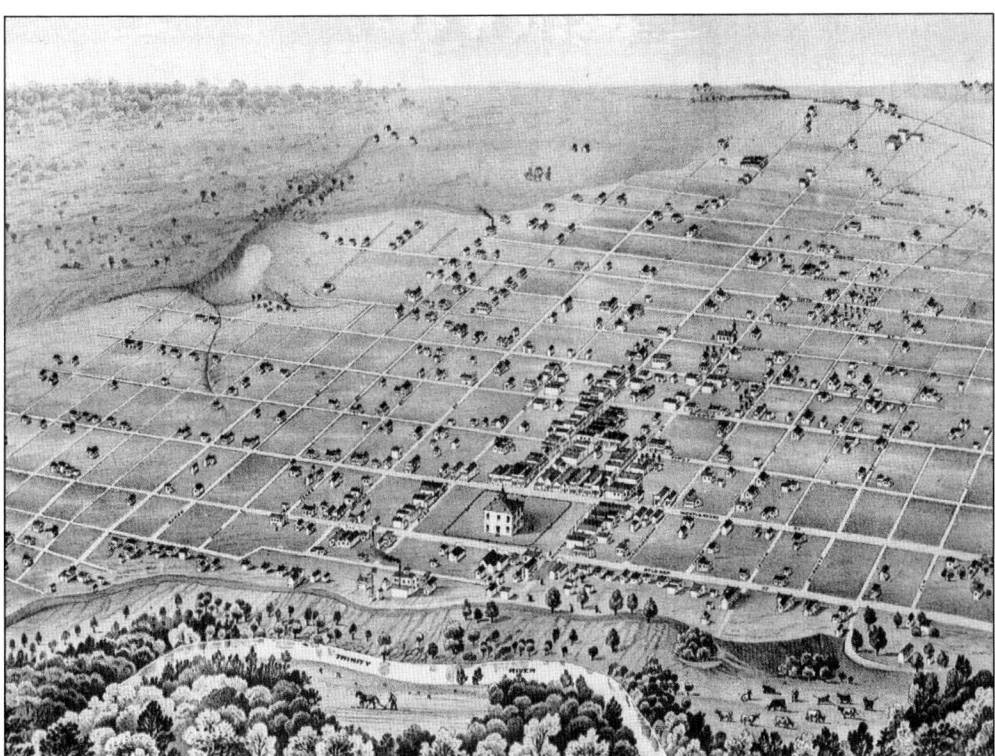

The northernmost outpost of a line of US Army forts built to protect the Texas frontier was named Fort Worth. This was no stockade, and the log buildings were arranged around a parade ground that later became the courthouse square. Cavalry stables were erected in the northern area of the post so that the horses could be led down the riverbank to be watered in the Trinity River. The fort was manned only from 1849 to 1853, by which time the frontier had moved west. But the Tarrant County Courthouse was built in the middle of the old parade ground in this bird's-eye view of Fort Worth. The community soon became known as Cowtown. (Courtesy Amon Carter Museum of Western Art, Fort Worth.)

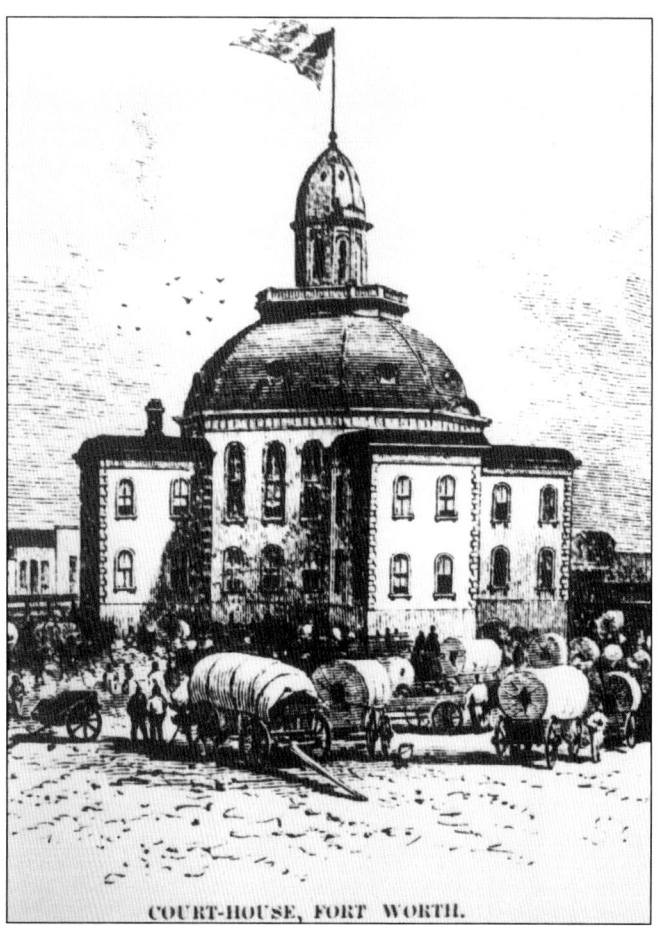

COURT-HOUSE, FORT WORTH.

The 1876 courthouse burned after just two years. The replacement, seen here, was built in 1878 and lasted into the 1890s, when a much larger building was needed. (Courtesy *Harper's New Monthly Magazine*, 1879.)

The ladies and children below stand northeast of the Trinity River, with the growing town of Fort Worth, known then as Cowtown, in the background. Note the 1878 courthouse skyline in the center of the photograph. (Courtesy Fort Worth Courthouse.)

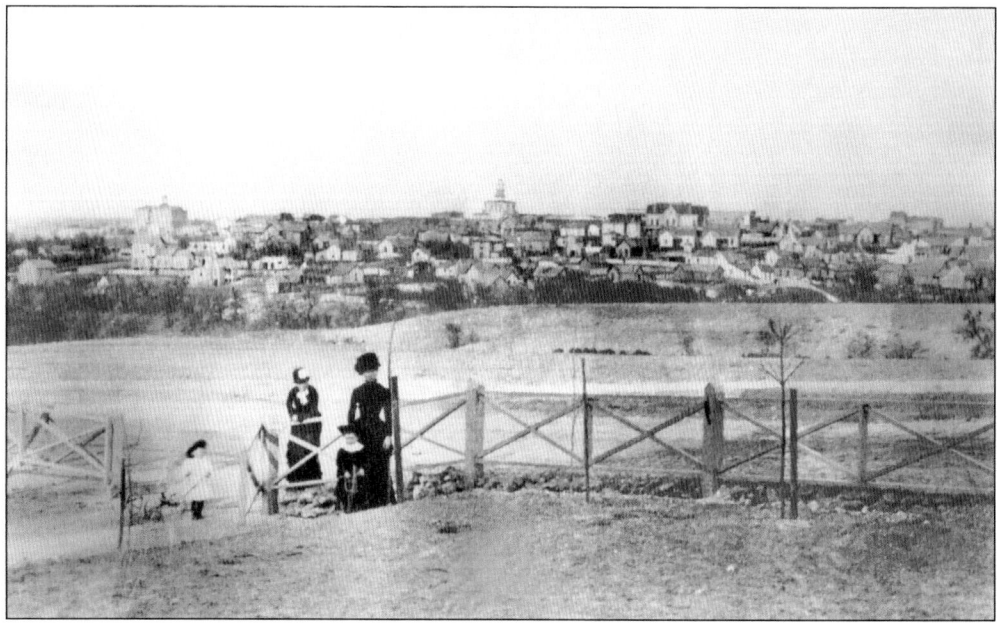

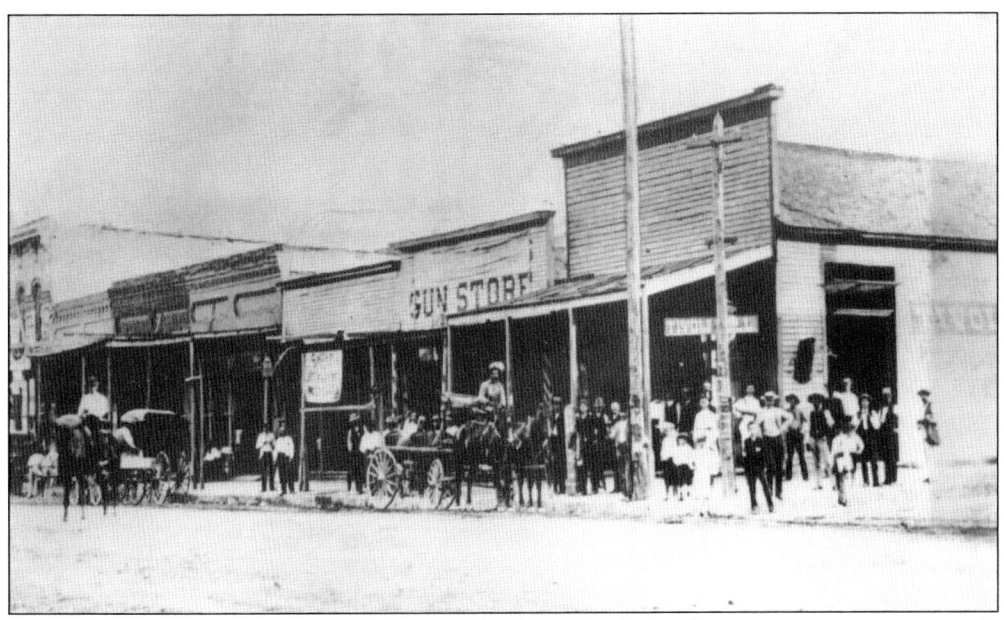

This 1876 street scene in Fort Worth features the false-front look of cow towns in the range cattle area. The Alamo Saloon on the square was one of numerous watering holes around town, and the red light district was known as "Hell's Half-Acre." (Courtesy Fort Worth Courthouse.)

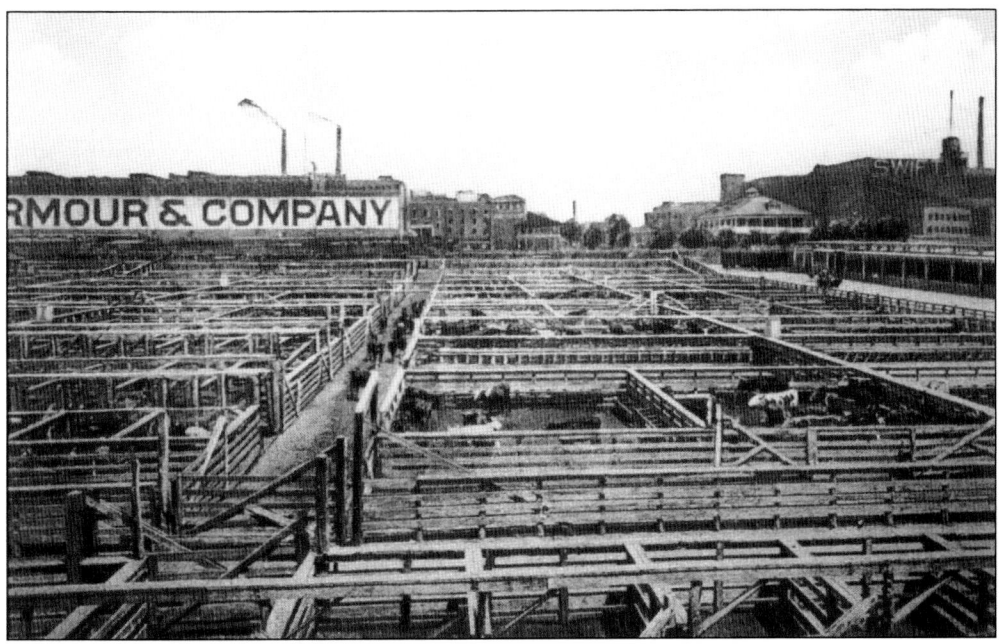

The Chisholm Trail passed through Fort Worth, and the first of several railroads reached "Cowtown" in 1876. The Fort Worth Stockyards grew into a major "Livestock Hotel," and in 1902, Armour and Company and the Swift Packing Company built packing plants near the stockyards. (Author's collection.)

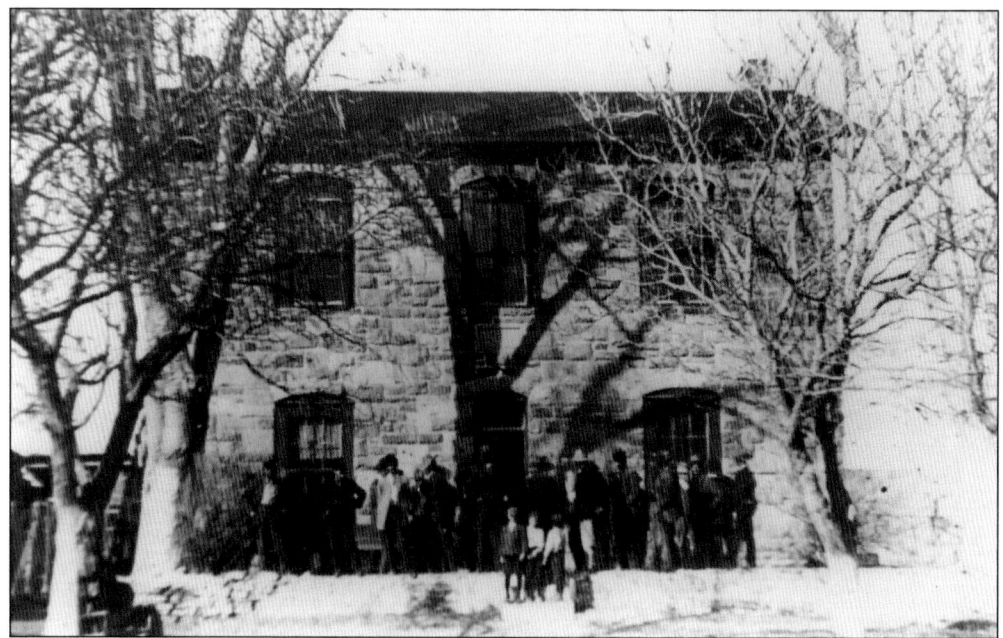

Founded in the Texas Panhandle in the mid-1870s, Tascosa was in the midst of a superb grazing area where great cattle ranches such as the LIT, the LS, the Frying Pan, and the enormous XIT prospered. In 1880, Tascosa became the seat of newly organized Oldham County, and a two-story rock courthouse was built, which still stands today. (Courtesy Panhandle-Plains Historical Museum, Canyon.)

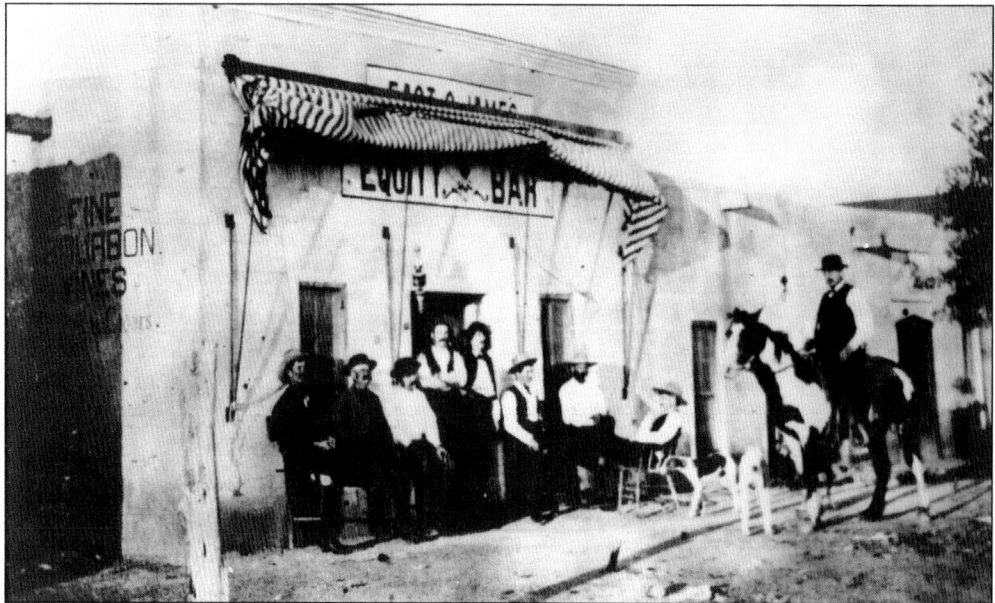

In the 1880s, Tascosa experienced its rowdy heyday as the "Cowboy Capital of the Panhandle." The first of the decade's 10 shoot-outs exploded in 1881 at the Equity Bar, when drunken trail boss Fred Leigh was shot dead by Sheriff Cape Willingham. Tascosa's last gunfight was also in the Equity, when Sheriff Jim East killed a gambler in 1890. (Courtesy Panhandle-Plains Historical Museum, Canyon.)

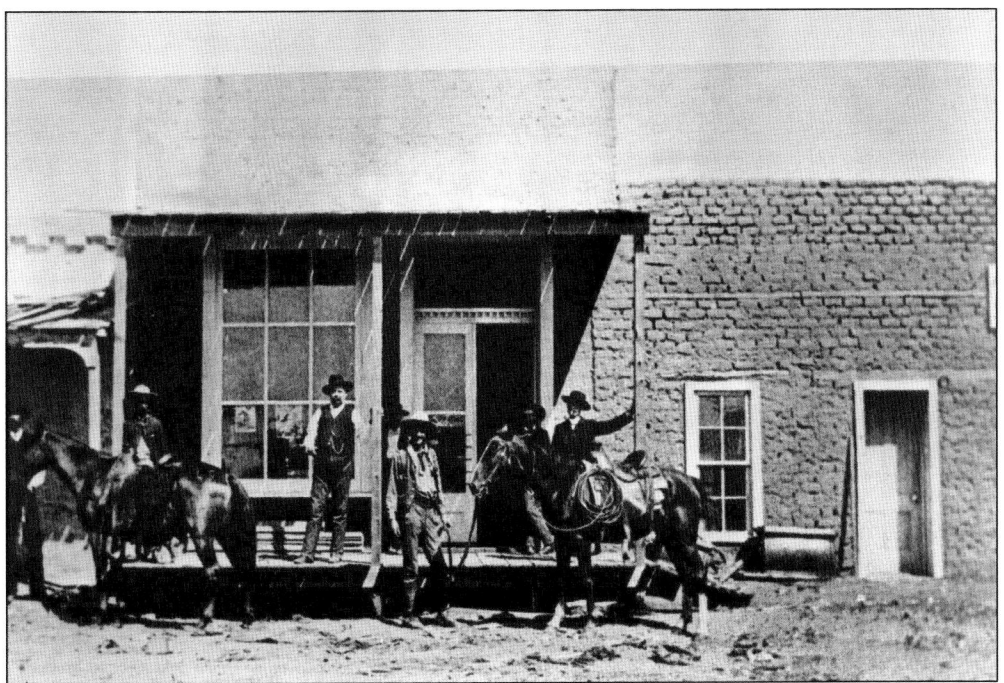

Wright and Farnsworth's general store was on the northeast corner of Tascosa's main intersection. There were four fatal gunfights in town in 1883, and in 1886, Tascosa's "Big Fight" saw four men killed in a saloon battle across the street from the general store. Three dead cowboys were laid out on the store's porch and covered with a tarpaulin before they were buried at Boot Hill. (Courtesy Panhandle-Plains Historical Museum, Canyon.)

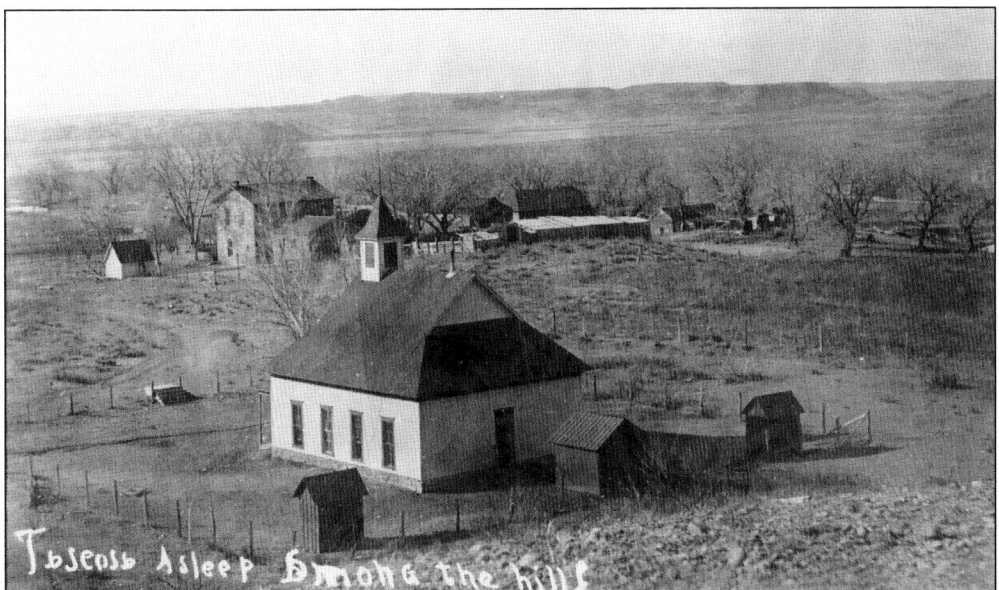

This view of Tascosa is from the north, with the rear of the two-story courthouse to the left of center. In the foreground is the one-room school, with separate outhouses for boys and girls. The population of the town reached 600 in the 1880s, but by the 1890s, Tascosa began a long decline. (Courtesy Panhandle-Plains Historical Museum, Canyon.)

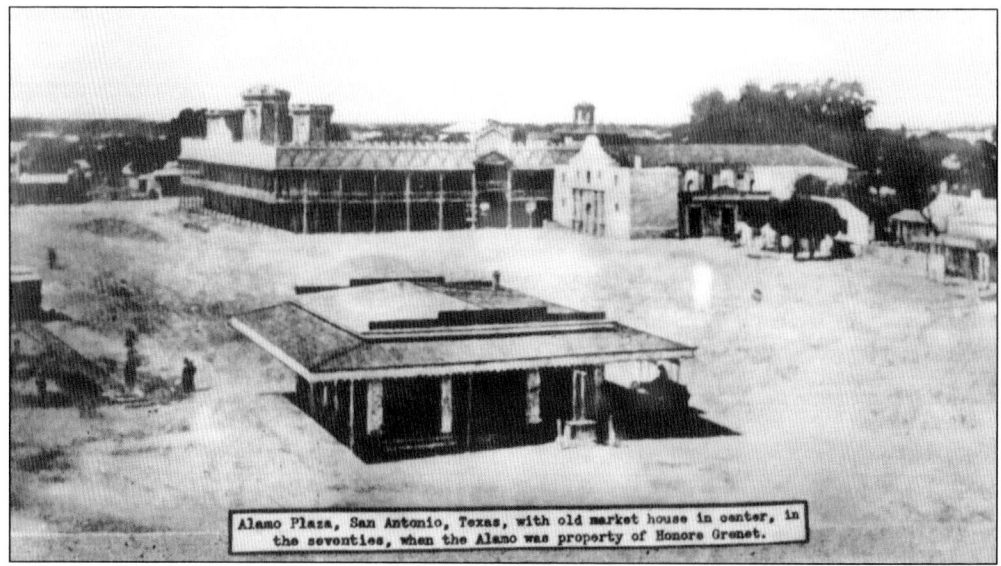

Alamo Plaza, San Antonio, Texas, with old market house in center, in the seventies, when the Alamo was property of Honore Grenet.

Throughout the 1700s, there were cattle herds around San Antonio from the five nearby missions. A century later, San Antonio was the starting point for cattle drives organized in south Texas. The town grew to include Alamo Plaza, where a circular barbed wire fence was temporarily erected to prove to cattlemen that "bobwire" could hold longhorns. (Author's collection.)

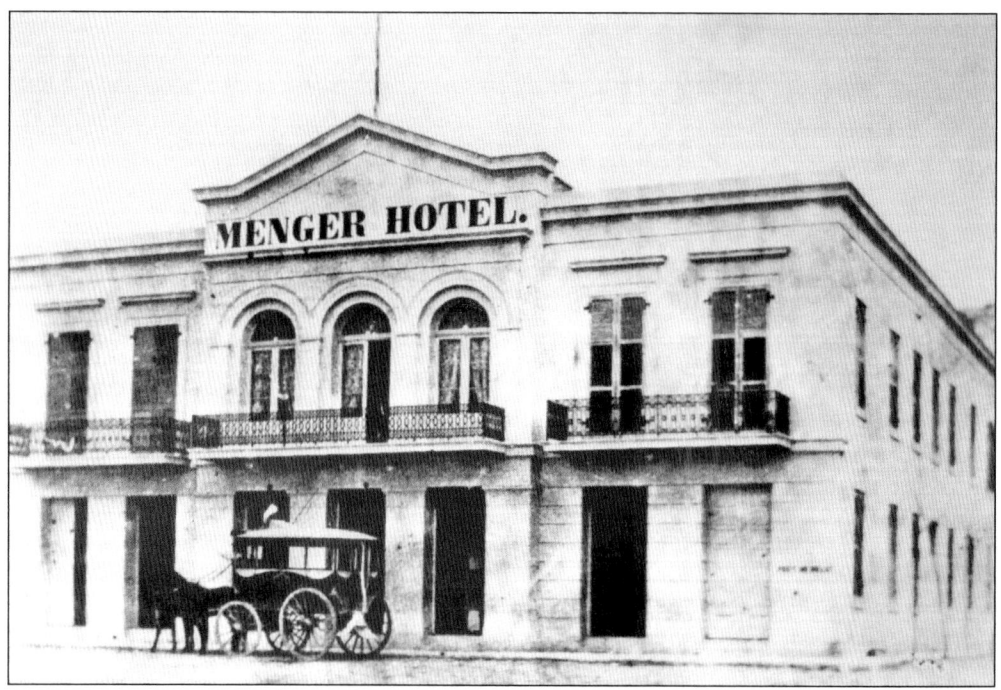

San Antonio's most popular cowboy saloon was the Buckhorn, while the favorite hostelry of cattlemen was the Menger Hotel, adjacent to Alamo Plaza. Richard King maintained a second-floor suite at the Menger for family trips to San Antonio and for ranch visitors. By the time he was 60, King had stomach cancer. He sought medical help in San Antonio but died at the Menger in 1885. (Author's collection.)

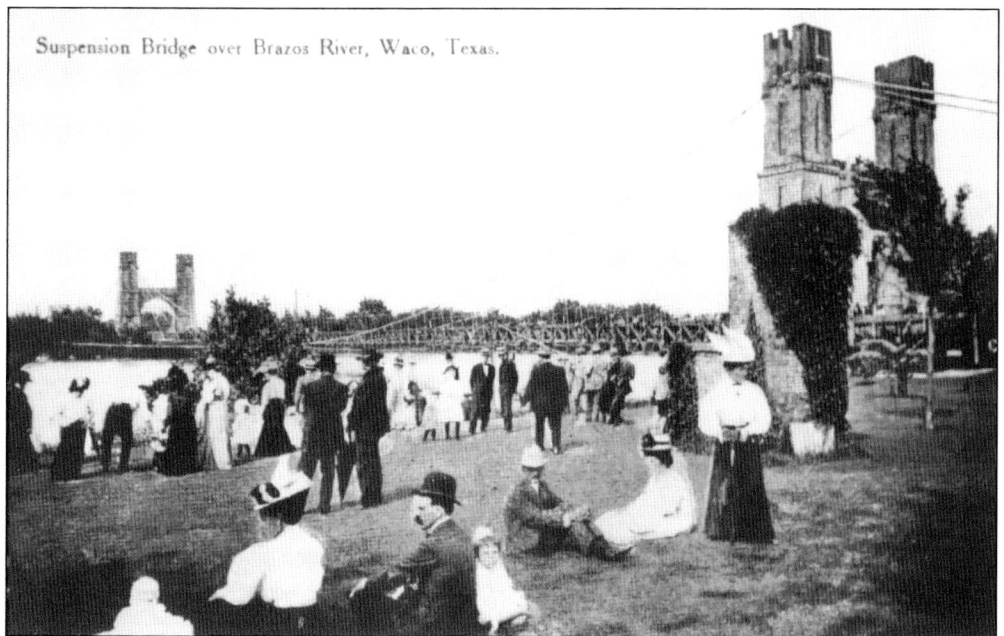

Many northbound herds feeding into the Chisholm Trail passed through Waco. There were saloons for cowboy recreation and enough gunfights that Waco was nicknamed "Six Shooter Junction." The Brazos River, just north of town, was more than 400 feet wide, and the Roebling Company was contracted to build what was at the time the world's longest suspension bridge. Many years later, the same company erected the much longer Brooklyn Bridge. (Author's collection.)

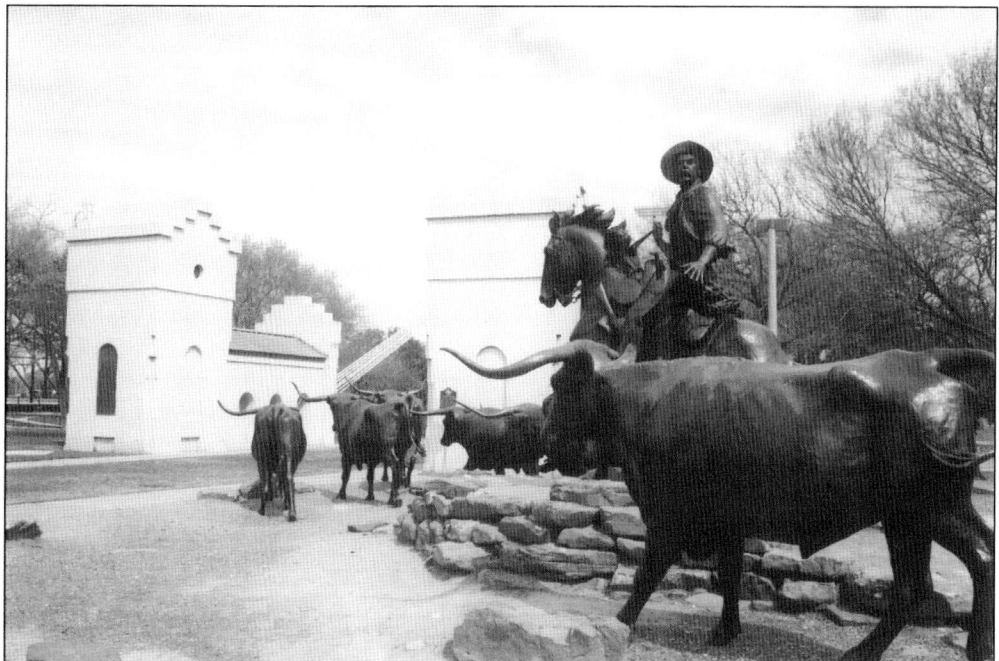

The Waco toll bridge was 475 feet long and was completed late in 1869 at a cost of $140,000. The city fathers who financed the bridge anticipated a stream of cattle herds, but many trail drivers found a ford somewhere and crossed there to avoid paying the toll. (Photograph by the author.)

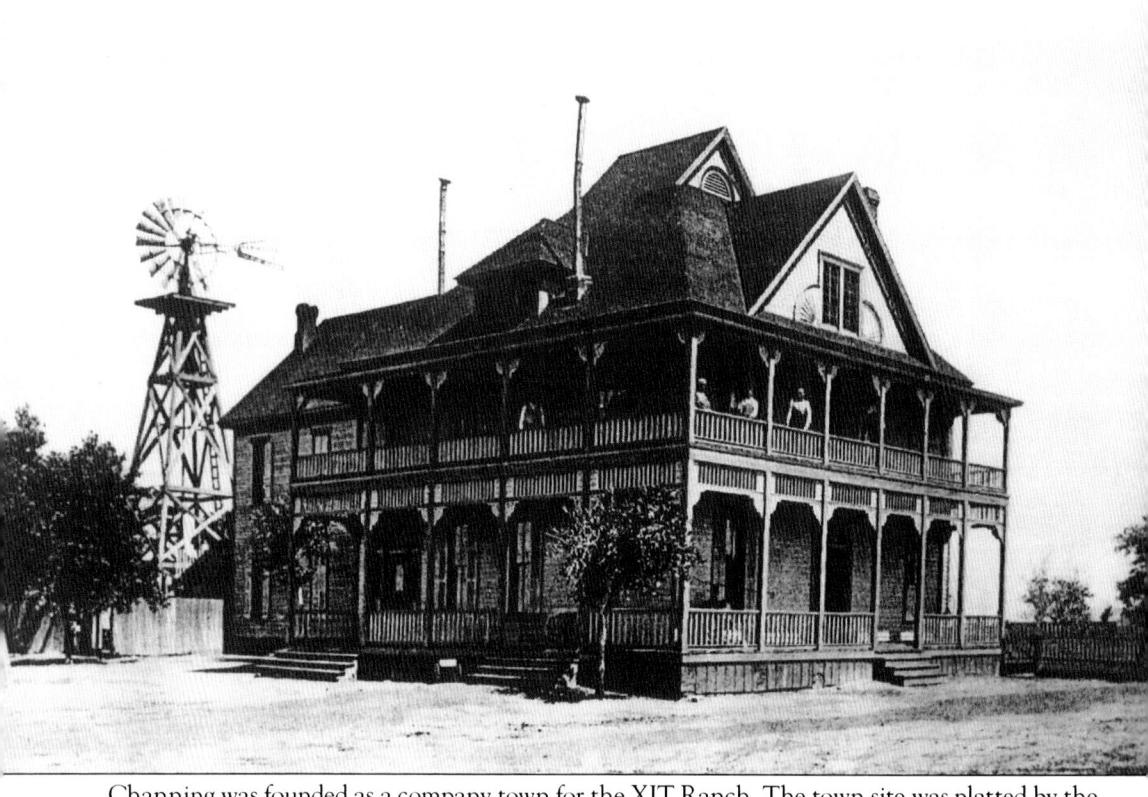

Channing was founded as a company town for the XIT Ranch. The town site was platted by the XIT in 1891 alongside the Fort Worth & Denver City Railroad. The first house was built by XIT general manager B.H. Campbell, and a brick XIT office was also built. Rita Blanco headquarters was only two miles south of town. Near the office was this two-story frame hotel; most of the customers were XIT visitors or executives. The hotel was the focal point of community Fourth of July celebrations. Each Christmas and New Year's, XIT cowboys hosted all-night dances at the hotel, providing turkey, deer, antelope, and, of course, beef. In 1903, the seat of Hartley County was moved from the hamlet of Hartley to Channing. XIT cowboys put the frame courthouse on wheels and towed it with horses to Channing. (Author's collection.)

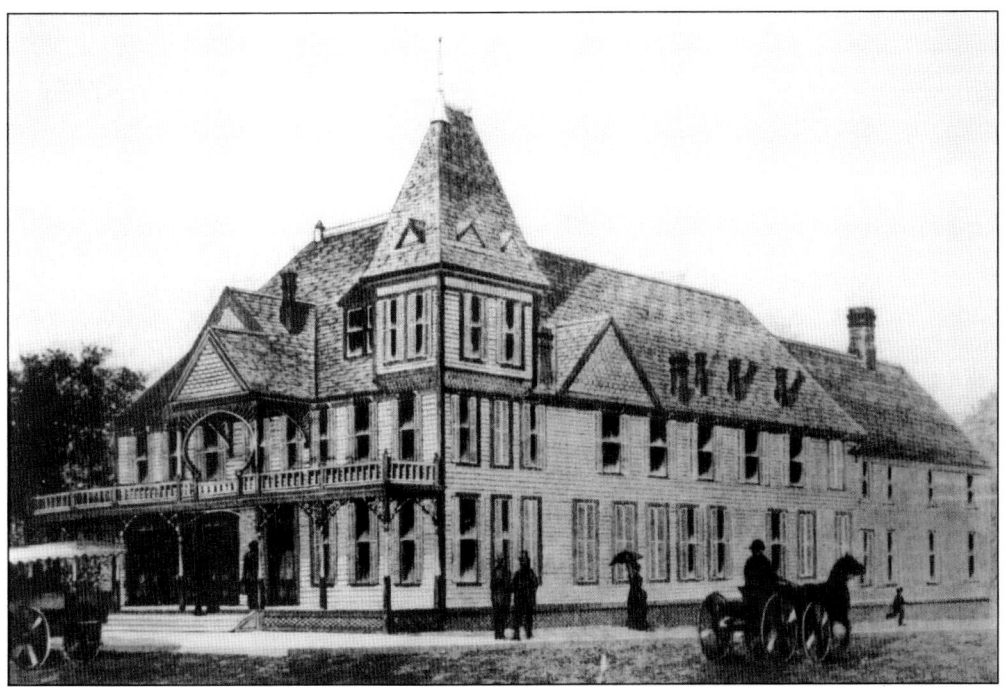

The 30-room Amarillo Hotel was built in 1889, only two years after Amarillo was founded. Harry B. Sanborn, a rancher and barbed wire promoter who was instrumental in founding Amarillo, built the hotel. (Author's collection.)

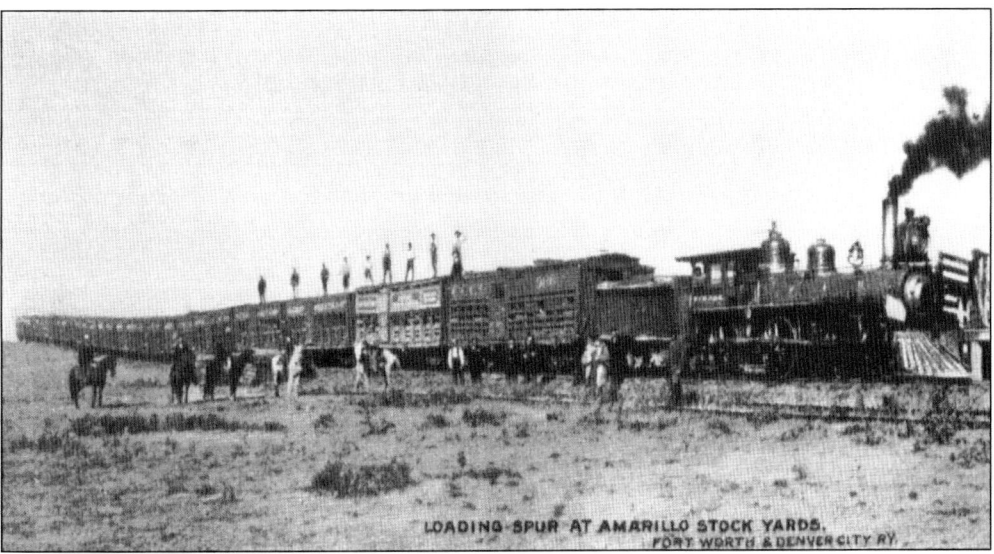

The arrival of the Fort Worth & Denver City Railroad in 1887 resulted in the founding of Amarillo in the middle of the Panhandle ranching country. Stockyards were built and Amarillo boomed as a cattle-shipping point. More railroads arrived, and Amarillo became one of the world's leading shipping points. In true cattle town style, a notorious area of saloons and brothels developed, known as the Bowery District. (Courtesy Historical and Biographical Record of the Cattle Industry in Texas.)

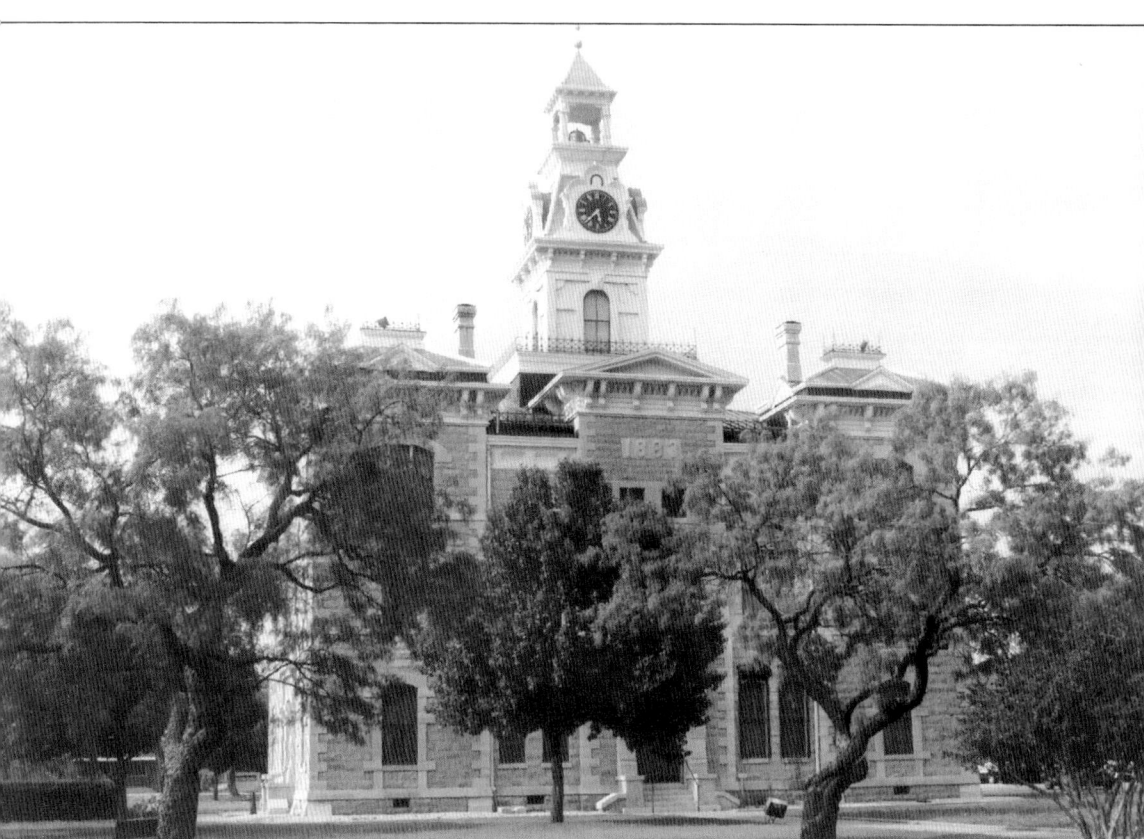

Albany was selected as the seat of the recently organized Shackleford County in 1874, several months before the town site was platted and lots were sold. The nearby Western Trail to Dodge City made the new town a supply center for herds and area ranches. When the Texas Central Railroad arrived in December 1881, Albany became a shipping point for cattle. In 1883, a substantial courthouse was built that, along with other structures from the late 1800s, continues to serve the community. By the early 1900s, Albany was nicknamed the "Home of the Hereford." (Photograph by the author.)

Seven
TROUBLE ON THE RANGE

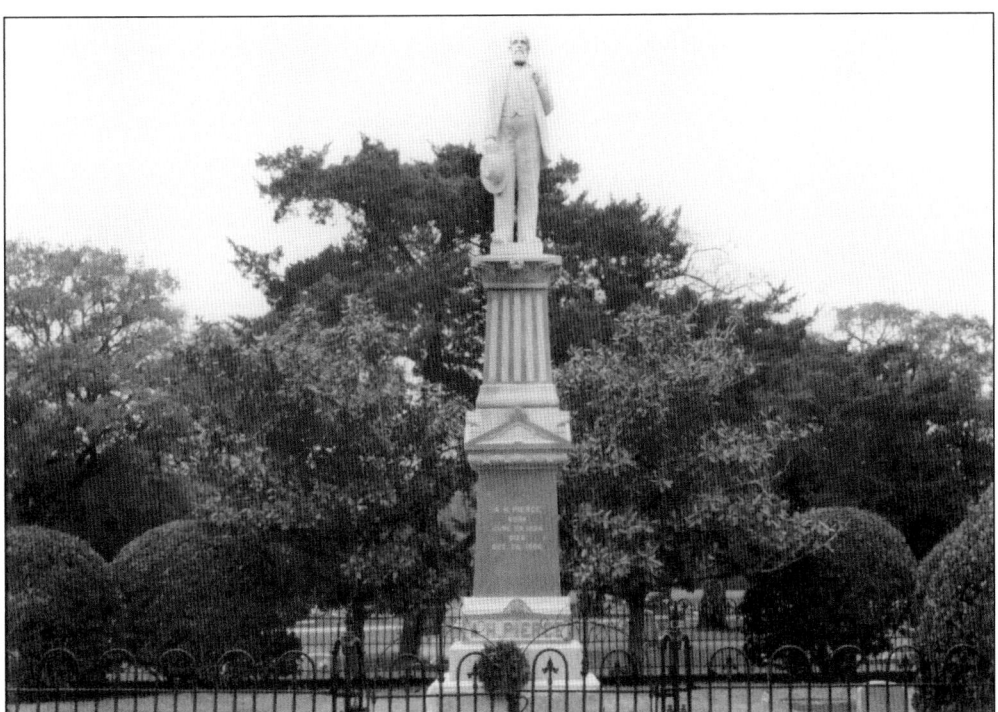

This towering statue marks the grave of Shanghai Pierce, but in 1871, Pierce sent five rustlers to their graves. With his vast herds frequently raided by stock thieves, Pierce learned that rustlers possessing green hides from stolen cattle were camped in a grove, so he sent riders to his camps and other ranches. Before dawn, he led 80 men into the grove, capturing five rustlers and hanging them from a dead tree. Legal action was commenced against Pierce, but he spent the next year and a half outside Texas buying and selling cattle until his brother wrote him that "atmospheric conditions" had cleared enough for him to return home. (Photograph by the author.)

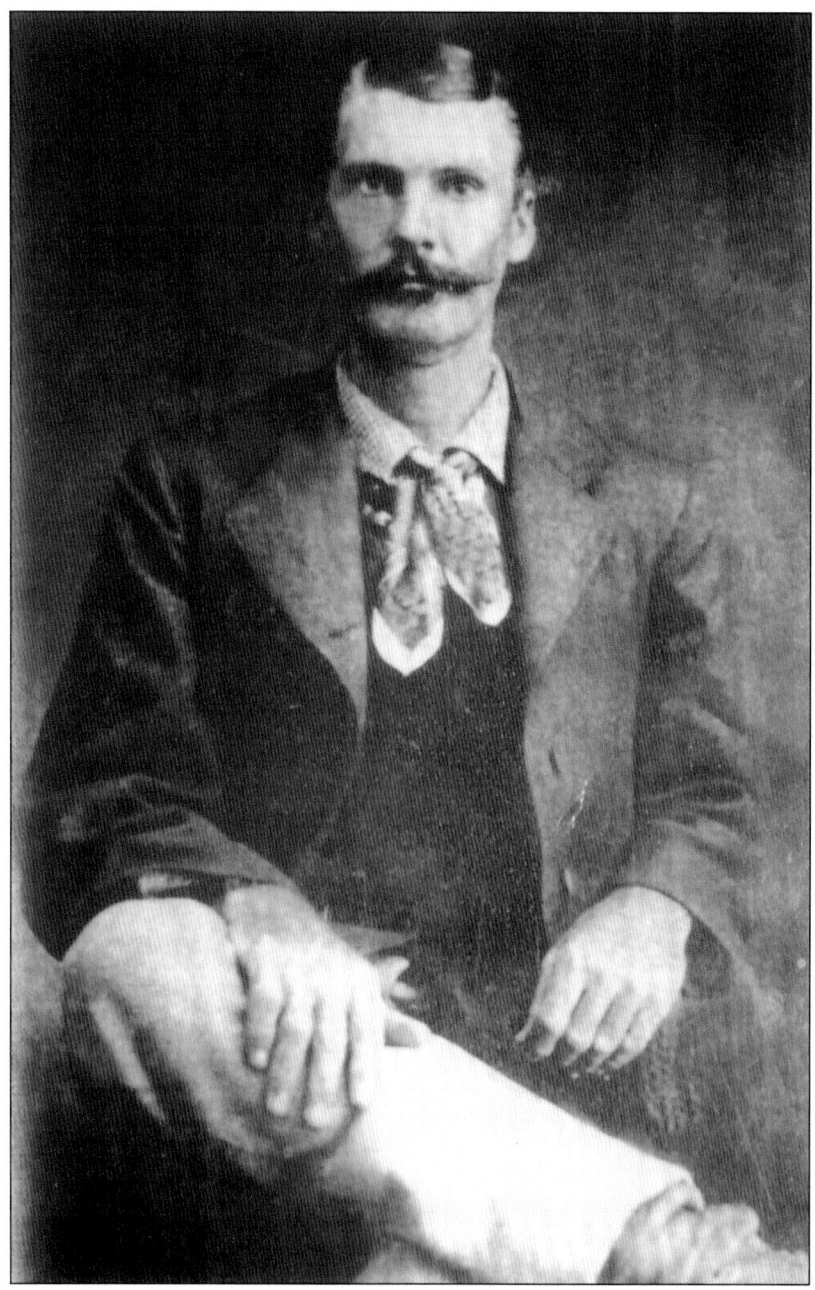

From the time he was a teenager in Lampasas County, Pink Higgins rode with pursuit parties against Comanche raiders and other stock thieves. In 1869, when he was 18, Higgins and several other members of the Law and Order League gave chase to a horse thief, doggedly following their prey for two days and two nights. Along the way, Law and Order League supporters provided fresh mounts. When the rustler was finally overtaken, the tired pursuers decided to apply summary justice. A rope was thrown over the limb of a hackberry tree, and Higgins adjusted the noose around the captive's neck. The rustler gamely allowed that he knew he was going to hell and that he wanted to arrive in time to find a partner for the first dance. Then he spurred his horse out from under himself and commenced his final journey. (Courtesy Betty M. Giddens.)

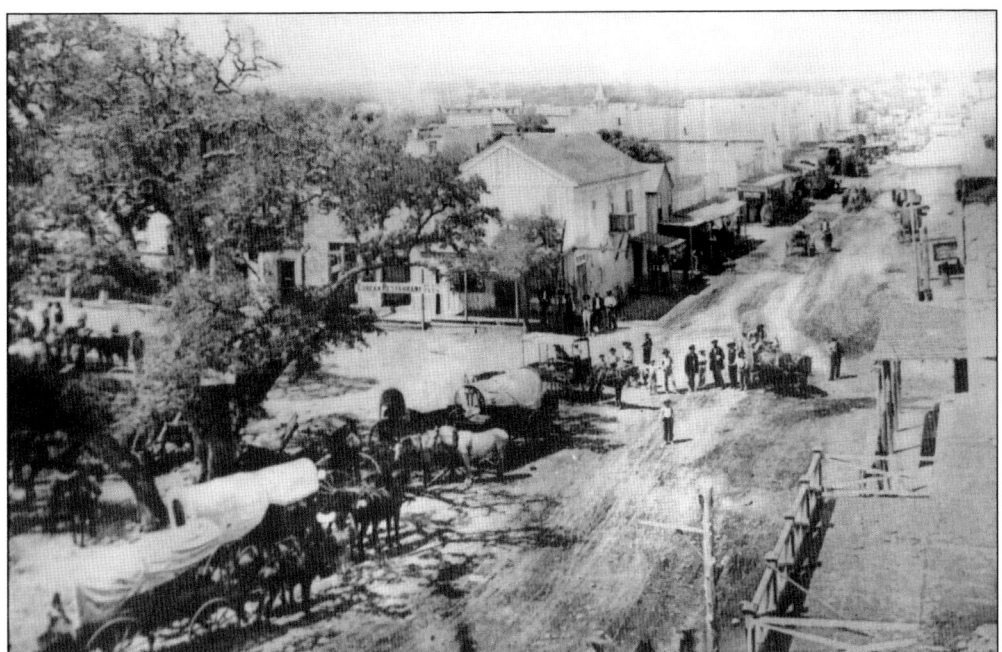

By the mid-1870s, Pink Higgins was an established rancher and trail boss. But the Horrell brothers, notorious stock thieves, began stealing his cattle. In a saloon on the west side of the Lampasas town square—obscured by a tree branch in this photograph—Pink killed Merritt Horrell with four Winchester rounds. The killing launched the Horrell-Higgins feud, which climaxed on June 7, 1877, with a gun battle on the square. (Courtesy Keystone Museum, Lampasas.)

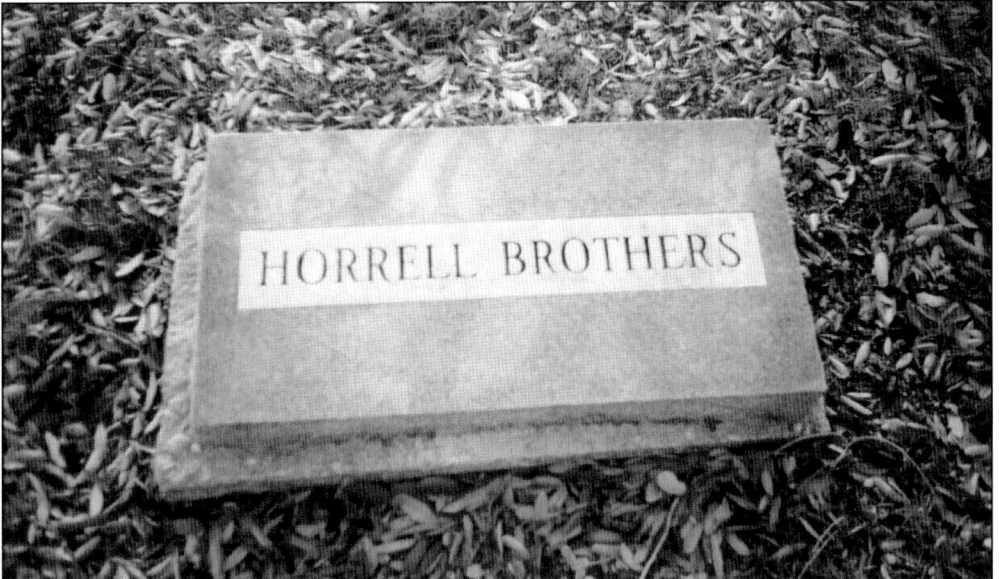

After the shootout on the square, Texas Rangers descended on Lampasas and arranged a truce between the two factions. But during the vengeful aftermath common to blood feuds, Mart and Tom Horrell were shot to death by a lynch mob inside the Meridian jail. Mart, Tom, and perhaps Merritt Horrell are buried in Oak Hill Cemetery in Lampasas. Two other Horrell brothers were shot to death in New Mexico. (Photograph by the author.)

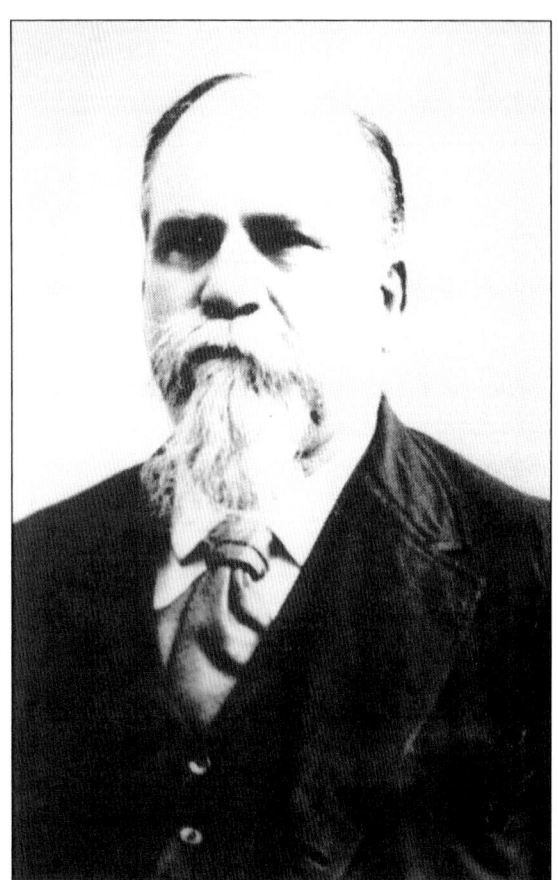

James C. Loving took over the cattle operation of his father, Oliver Loving, after he was killed in 1867. Like other cattlemen on the Texas frontier, J.C. Loving was plagued by stock thieves. In 1877, he was instrumental in organizing a meeting in Graham where 40 ranchers founded the Stock-Raisers Association of Northwest Texas. Loving served as secretary of the organization for 27 years until his death in 1902. (Courtesy Historical and Biographical Record of the Cattle Industry in Texas.)

The Stock-Raisers Association of Northwest Texas rapidly grew in membership and effectiveness, with name changes to the Cattle Raisers Association of Texas and then to the Texas and Southwestern Cattle Raisers Association (SWCRA). Scornful of the relatively mild term "rustlers," association members have long preferred "damn cattle thieves." The old brand book (below) is on display at the SWCRA Museum. (Photograph by the author.)

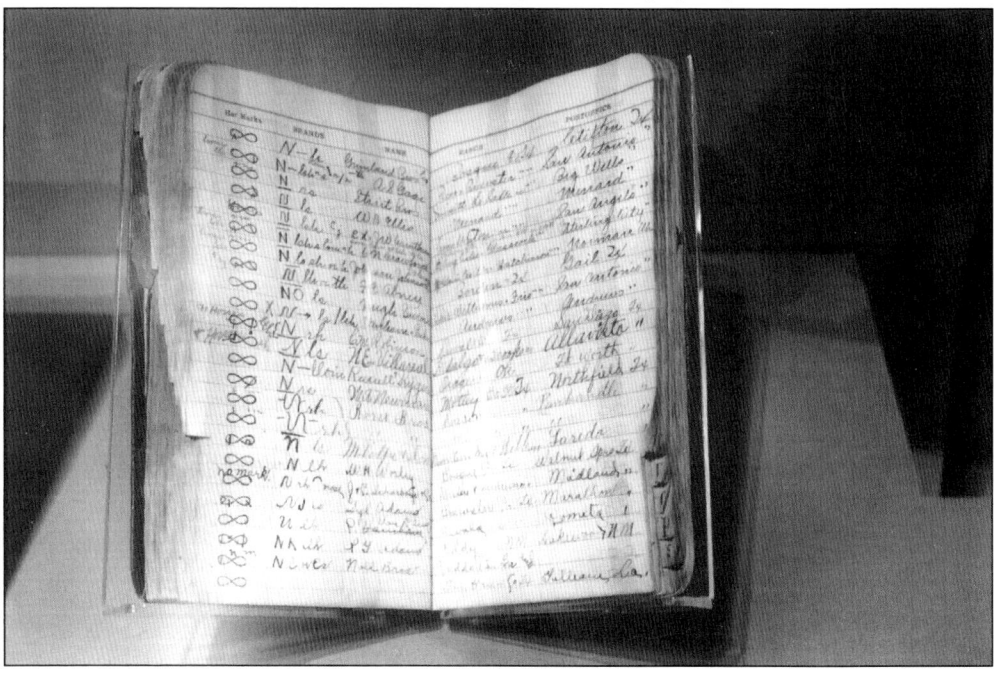

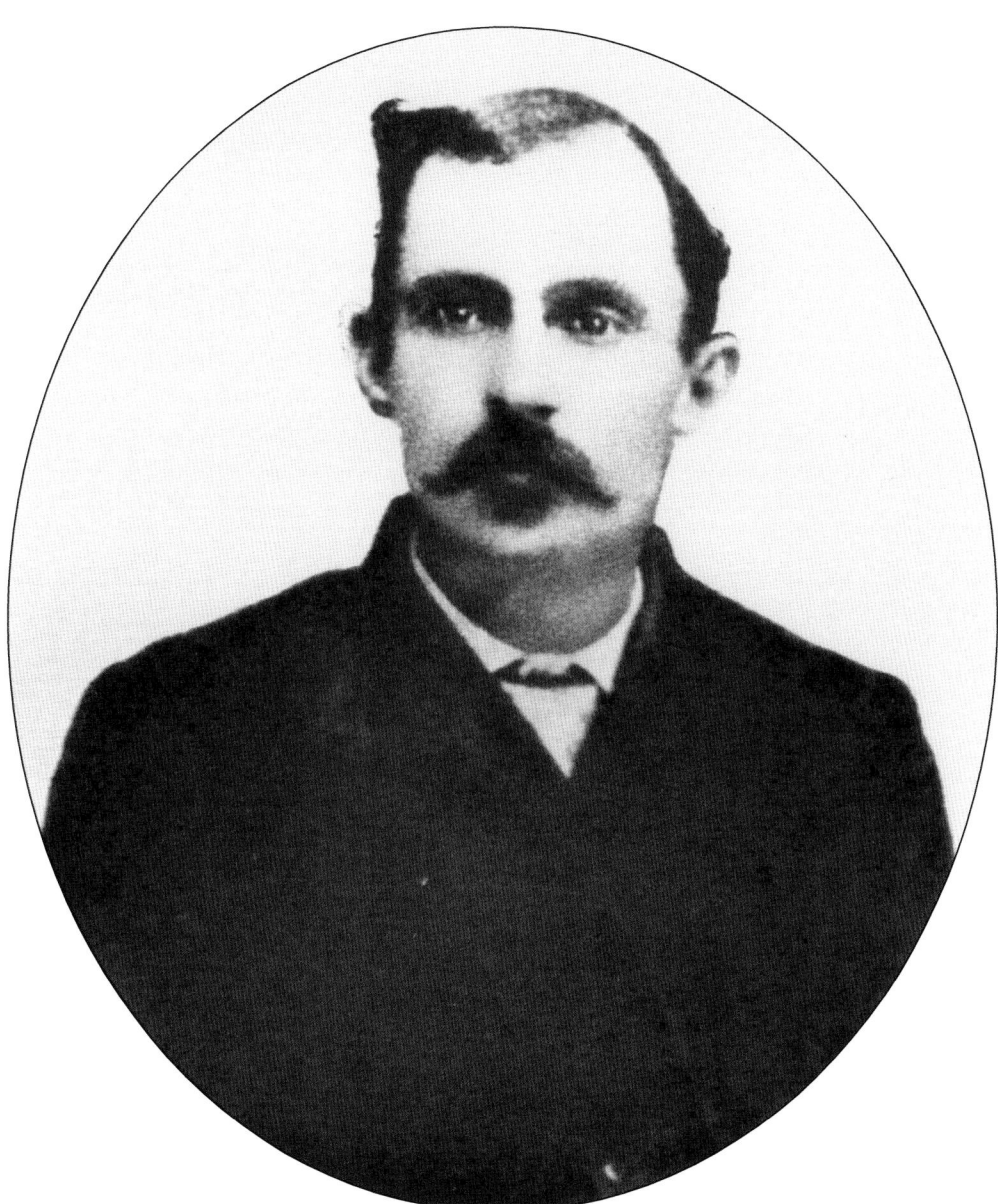

James B. Miller, the West's premier assassin, was called "Killin' Jim" and "Killer Miller," although not to his face. He played a role in the range conflict between cattlemen and sheepherders. Across the West, there were nearly 130 violent incidents between these factions. At least 28 sheep men and 16 cowboys were killed, and more than 53,000 sheep were shot, clubbed, knifed, poisoned, dynamited, and rim-rocked. Reportedly, Jim Miller was hired by Texas cattlemen at $150 per killing. "I have killed eleven men that I know about," he admitted to a Fort Worth acquaintance, before adding with disdain, "I have lost my notch stick on sheepherders I've killed out on the border." (Author's collection.)

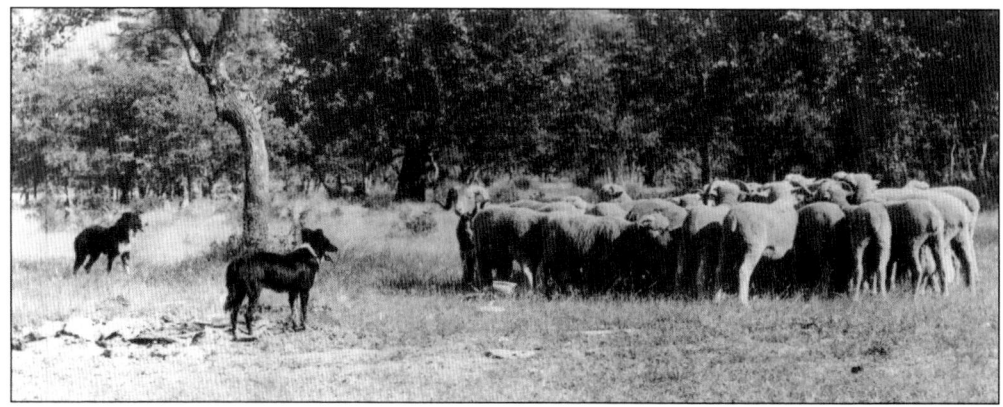

Sheep were cursed by cowmen as "hoofed locusts," "stinkers," "maggots," and "baa-a-ahs." Cowboys liked to say, "There ain't nothing dumber then sheep except a man who herds 'em." Cowmen also derisively referred to sheepherders as "mutton punchers" and "lamb lickers." (Author's collection.)

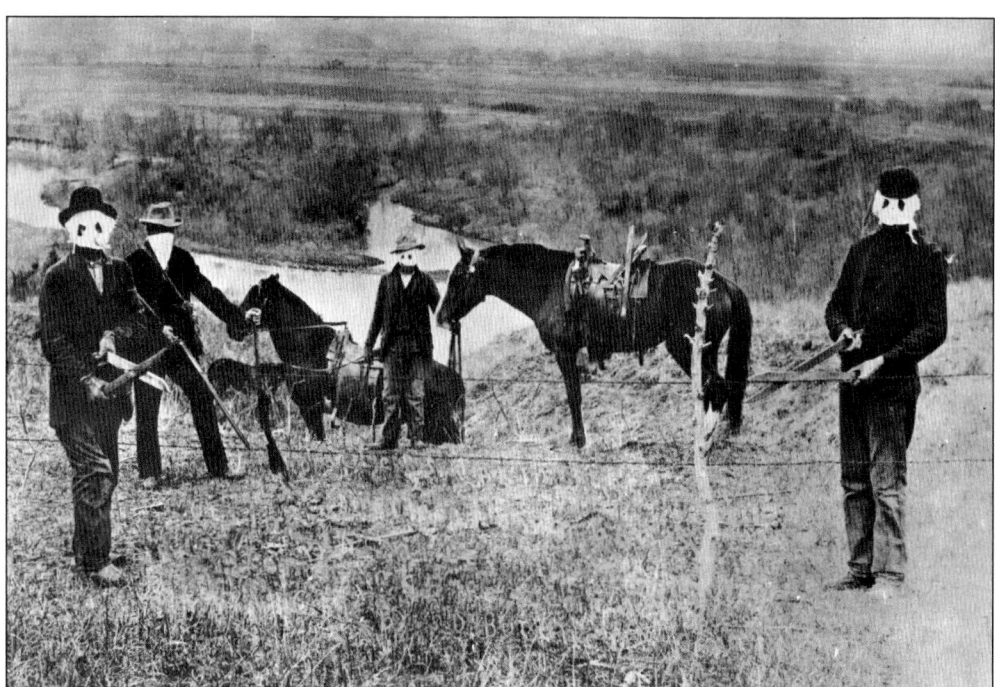

By 1883, there were fence-cutting episodes in half of the counties of Texas—although this photograph was taken in Nebraska—and much of the focus was against sheep pastures. Sheepman G.W. Mahoney of Coleman County, for example, suffered miles of cut fence, with the wire snipped twice between each fence post. (Courtesy Panola College Library, Carthage.)

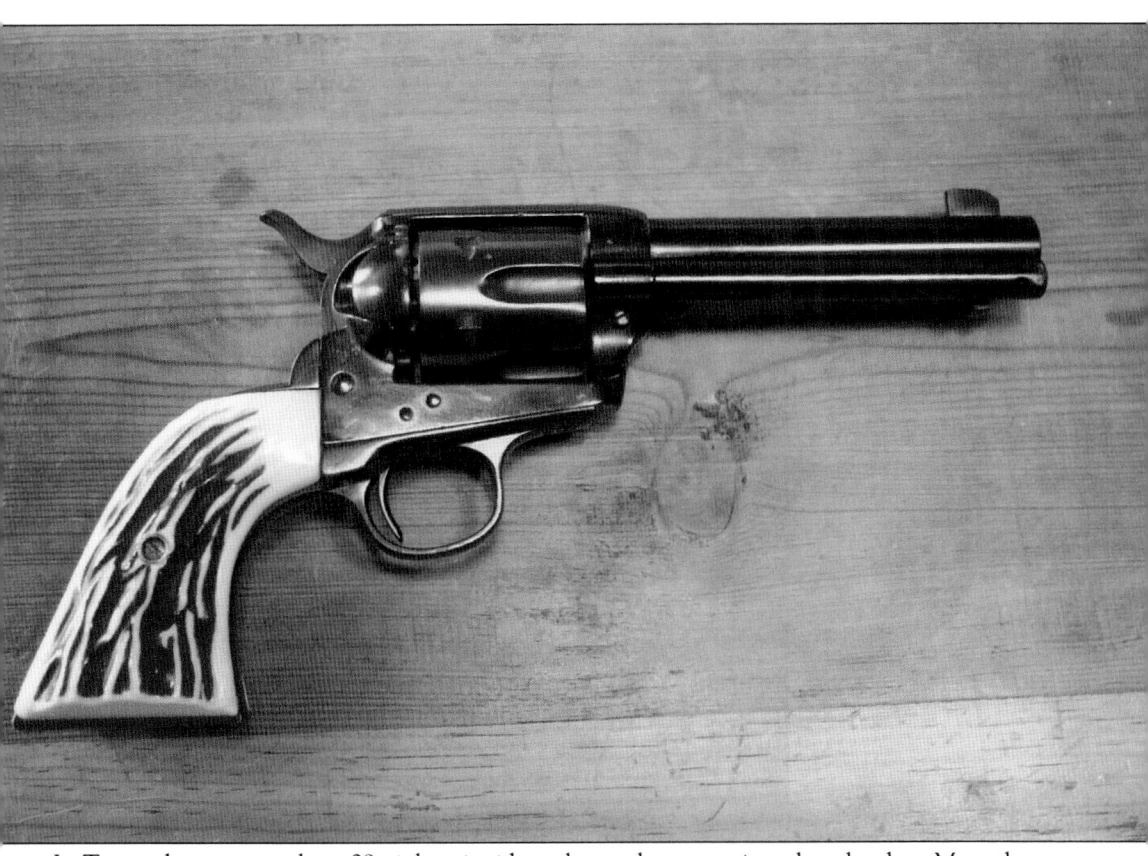

In Texas, there were at least 29 violent incidents by cattlemen against sheepherders. More than 3,200 sheep were slain, along with four sheepherders and two cowboys. In 1889, an out-of-work cowboy persuaded Burnet County sheep rancher Andy Feild to hire him as a herder. But the cowboy ignored his duties and was fired. Before leaving the ranch, the cowboy rode up to accost Feild. Dismounting, he cursed Feild, then drew his revolver and snapped off a shot. But Feild had taken the precaution of placing the Colt seen here in his waistband. Feild charged, shooting the cowboy in the elbow, chest, and head. When the Feild sheepdog licked the wounds, the corpse moaned but soon died. (Courtesy Andy Feild.)

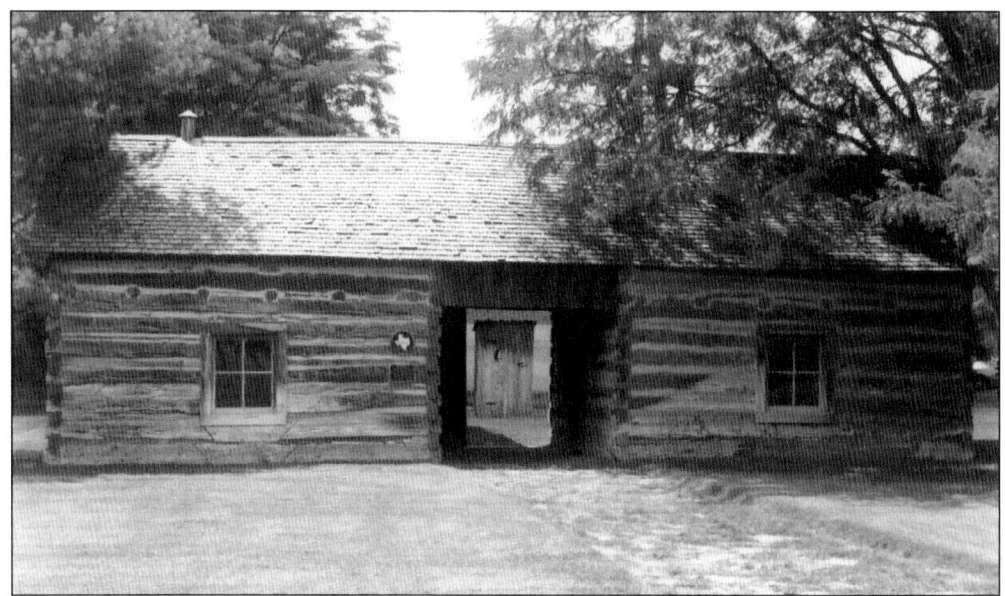

The T-Anchor ranch house was the first log cabin erected on the High Plains of Texas and became the site of decisive action in the rancorous cowboy strike of 1883. The T-Anchor, located north of present-day Canyon, put up the first big fence in the Panhandle, enclosing 240,000 acres of grass. Jule Gunter served as ranch manager in the spring of 1883, when Panhandle cowboys attempted to organize and strike for higher wages. Several T-Anchor men quit the ranch the day before the April 1 strike deadline, and Gunter was told that a cowboy delegation would ride to ranch headquarters to present demands. Gunter forted up with loyal hands inside the cabin and angrily sent the strike delegation away. The strike was a complete failure, as Gunter and other ranch managers easily found replacements. The T-Anchor buildings are now on display at the Panhandle-Plains Museum. (Photographs by the author.)

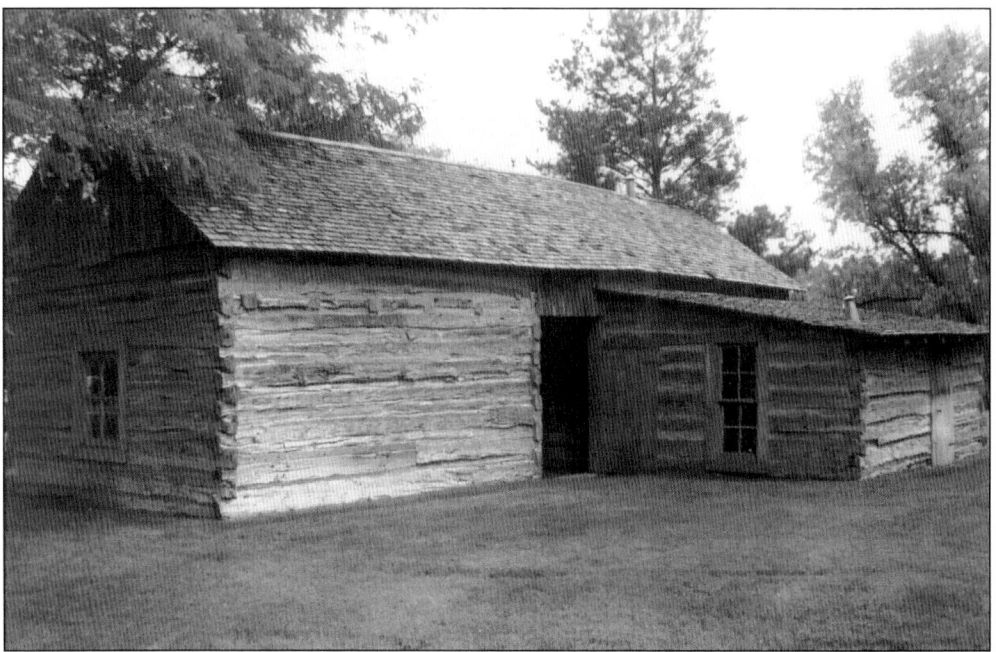

A major problem on the Texas range was the enormous network of prairie dog holes. By the late 1800s, there were an estimated 800 million prairie dogs in West Texas. Half were in a vast prairie dog town that spread across 25,000 square miles, stretching 250 miles north to south from Childress to San Angelo and approximately 100 miles east to west. (Photograph by the author.)

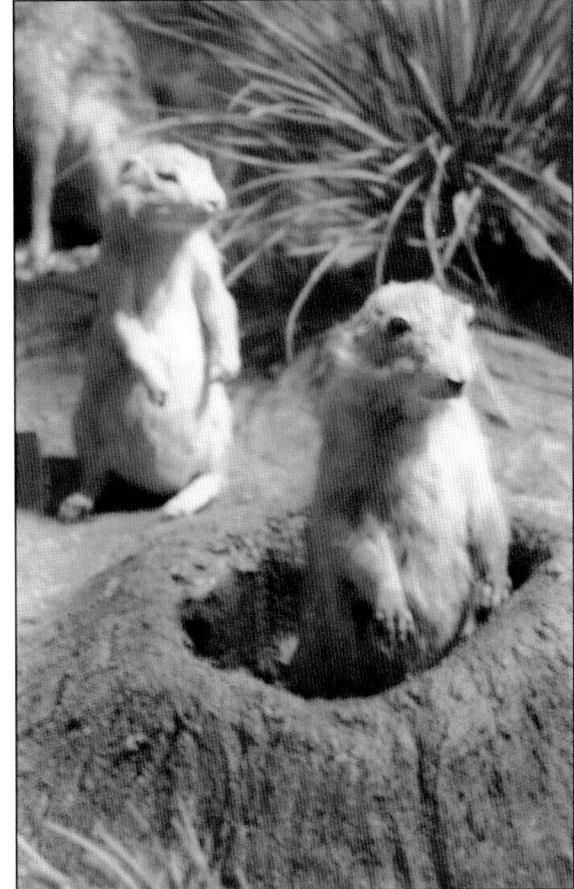

Prairie dogs ate and destroyed enough grass to feed more than three million cattle. Furthermore, cowboys were in danger if their horse stepped in a prairie dog hole. Famed range photographer Erwin E. Smith shot the scene below, probably posed, of cowboys riding to the aid of an unhorsed comrade. (Courtesy Panola College Library, Carthage.)

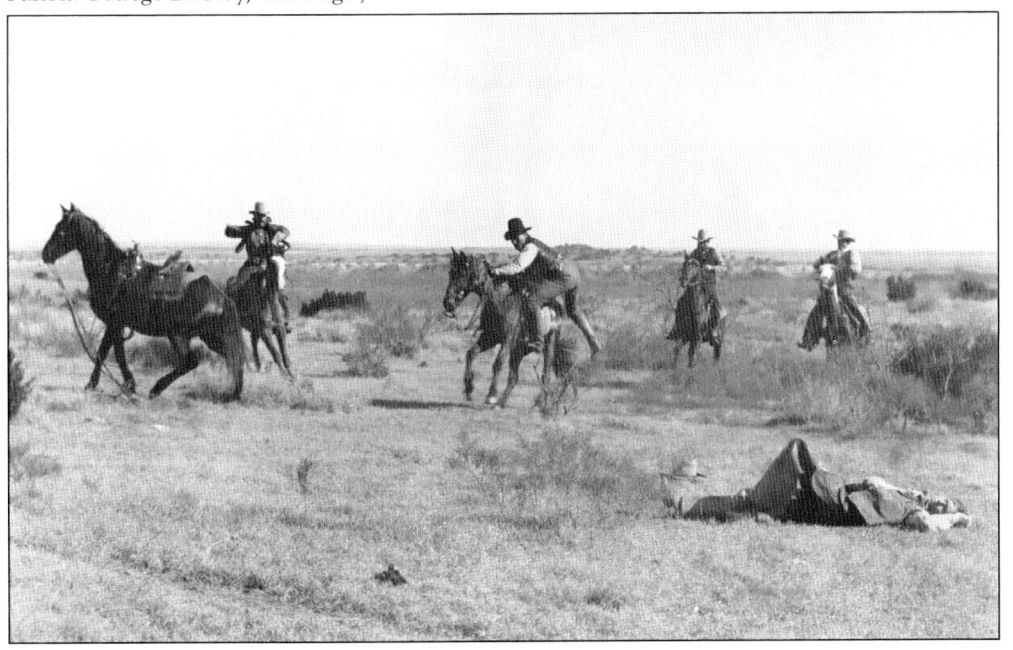

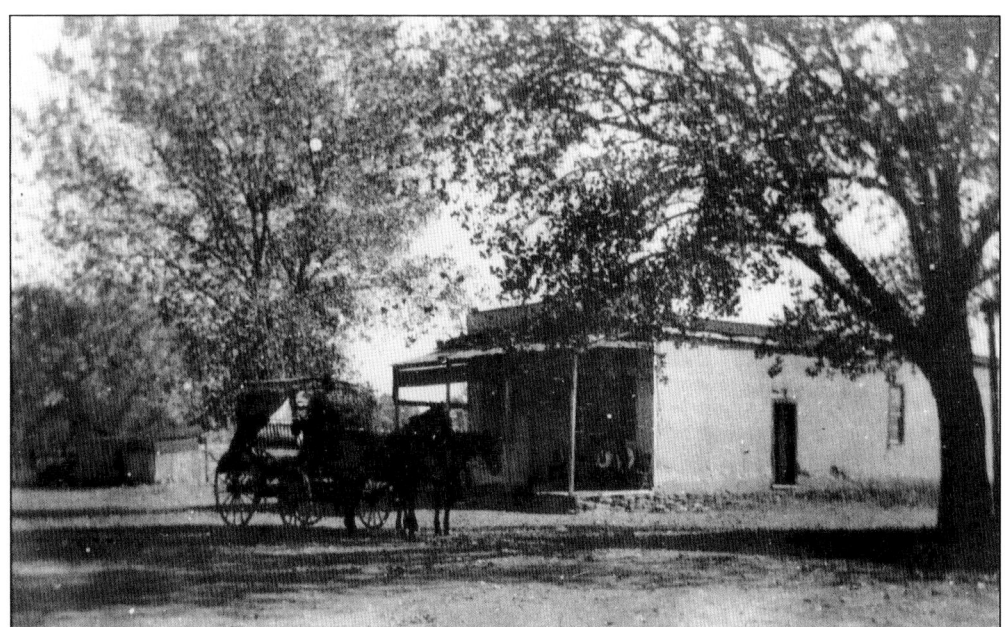

Tascosa's deadliest explosion of violence was an 1886 gun battle known as "The Big Fight." There were hard feelings against LS cowboys, and a post-midnight revelry resulted in an ambush at Tascosa's main intersection. One LS man was shot dead in the street from the Jenkins and Dunn Saloon. His three comrades charged into the saloon, and two of them were killed, along with a bystander. (Courtesy Panhandle-Plains Historical Museum, Canyon.)

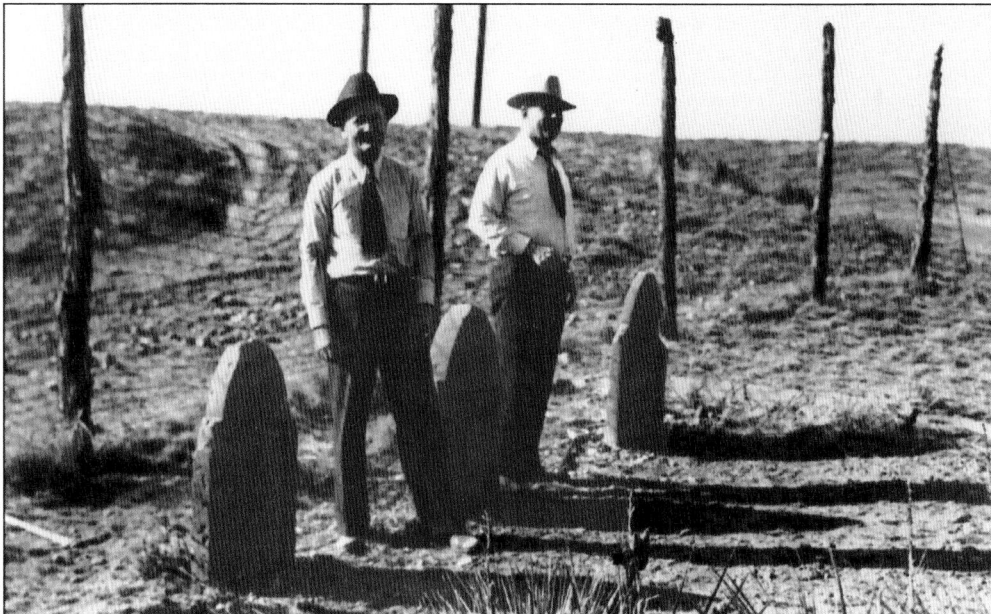

The next day, four coffins were built and the corpses were clad in new black suits. A mass funeral was held in the afternoon, and the entire populace of Tascosa, along with area cowboys, formed a half-mile procession to Boot Hill. In 1939, rancher, author, and professor J. Frank Dobie (right) and a friend examined the three LS tombstones. (Courtesy Panhandle-Plains Historical Museum, Canyon.)

Pink Higgins was hired as a stock detective by the vast Spur Ranch, which was constantly raided by rustlers. With Higgins's towering reputation as a gunman, several known rustlers soon moved to New Mexico, and the "cattle leakage" of the Spurs rapidly declined. Higgins built a board-and-batten house for his family on a small Kent County spread. (Author's collection.)

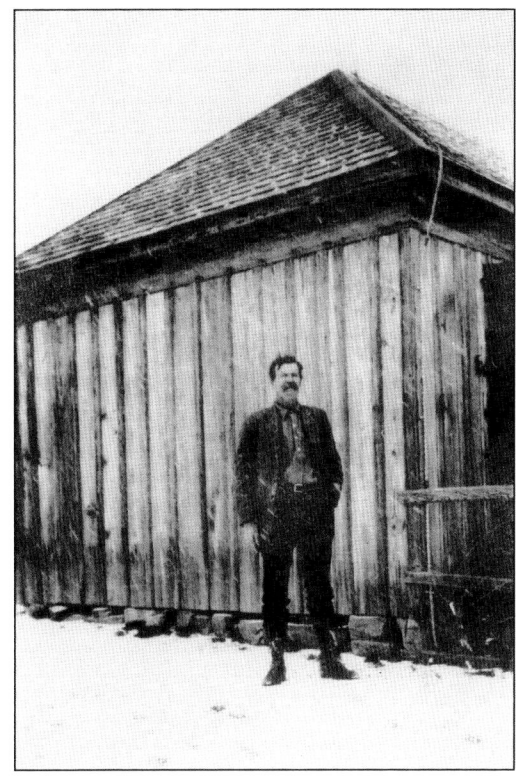

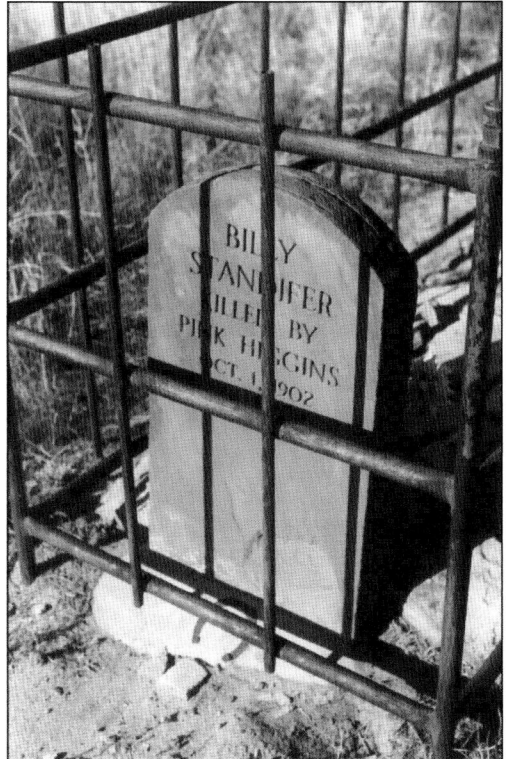

Billy Standifer, a former sheriff, also rode for the Spurs as a stock detective. But Standifer was a native of Lampasas County and entertained an enmity toward Higgins. There was a challenge, and on October 1, 1902, Higgins and Standifer engaged in a rifle duel near the Higgins home. Standifer was buried where he fell. (Photograph by the author.)

Billy Johnson was a teenaged drover on an 1878 cattle drive when he discovered the spring-fed Ennis Creek on the grasslands of frontier Scurry County. He placed a few head of his own cattle on the range and soon returned to build a dugout beside the creek. Johnson went on to accumulate 47 contiguous sections of land and prospered as a cattleman. (Courtesy Betty M. Giddens.)

Billy Johnson married and raised three sons and a daughter on his ranch. He built the two-story frame ranch house below, where his family posed on the porch. Note the well in the yard. (Courtesy Scurry County Historical Museum, Snyder.)

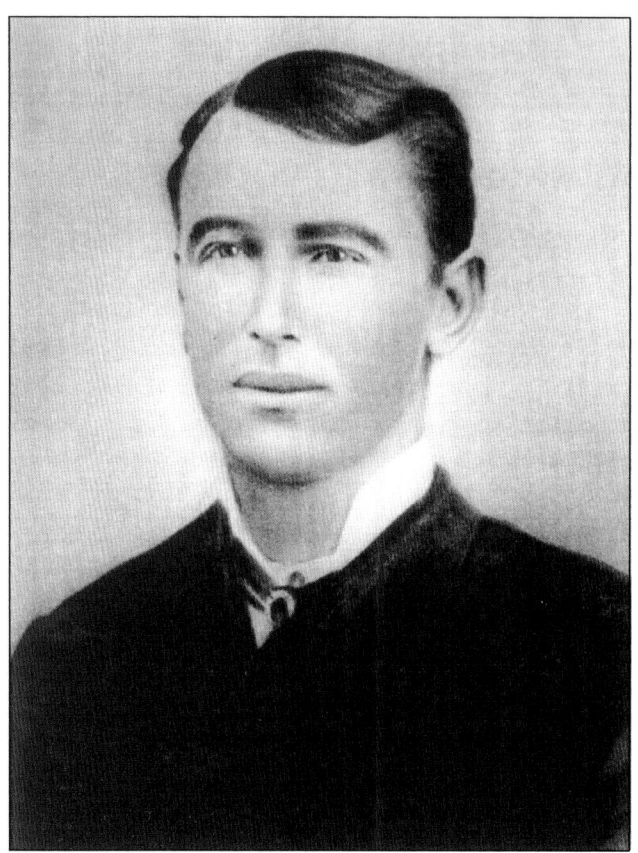

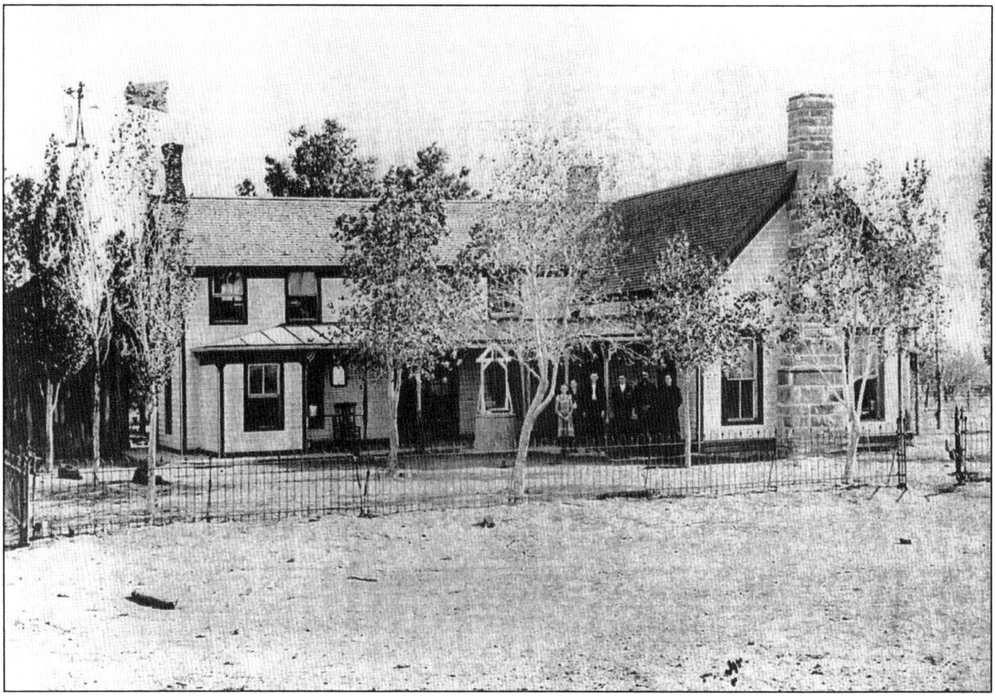

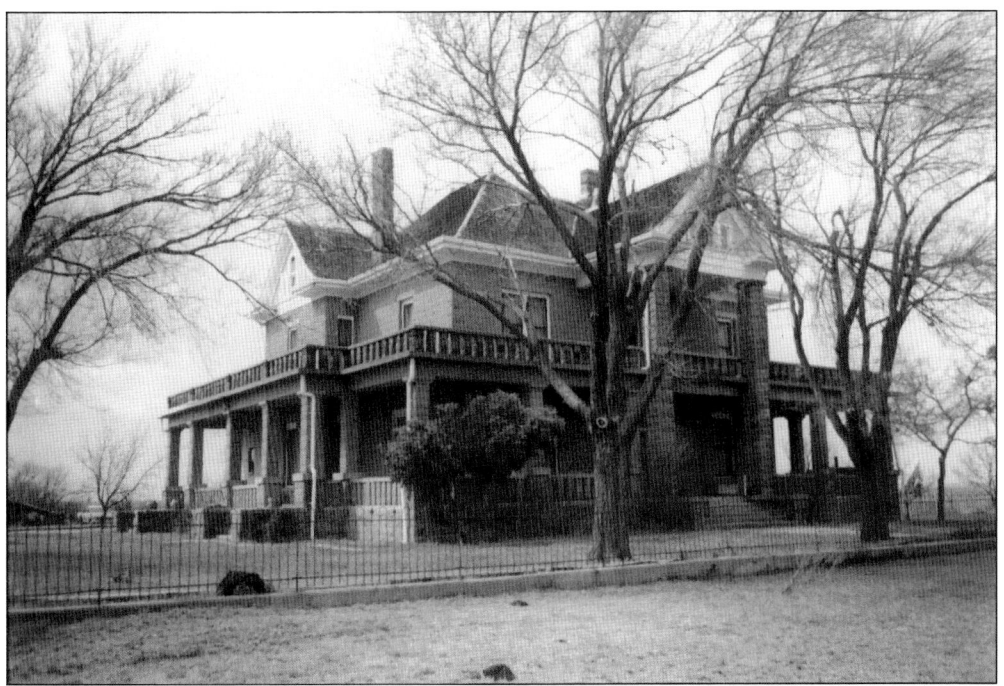

Successful as a rancher and businessman, Billy Johnson replaced his ranch house in 1910 with a 16-room mansion built of concrete blocks. The splendid old home still stands 12 miles north of Snyder. (Photograph by the author.)

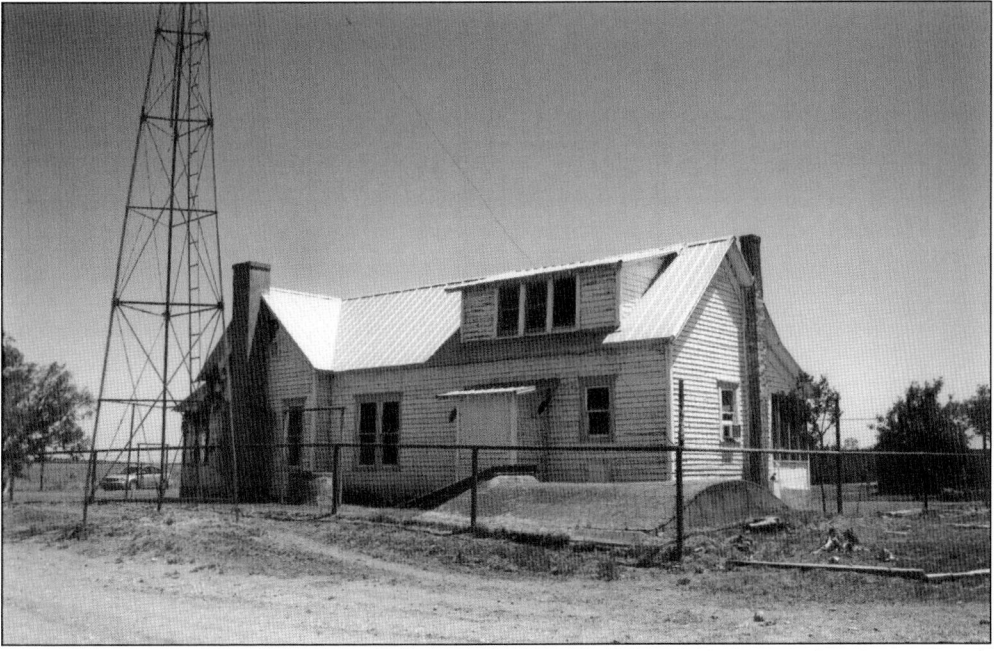

In 1887, Dave and Laura Belle Sims brought their cattle herd and growing family to rugged, isolated Kent County. Dave acquired 40 contiguous sections totaling more than 25,000 acres. The family lived in a dugout until a frame ranch house was built. Soon, there were 10 Sims children. (Photograph by the author.)

The Sims ranch was only 15 miles north of the Johnson ranch. In 1905, Ed Sims, 21, married 14-year-old Gladys Johnson. The couple had two daughters, but instead of a happy union of two ranching families, the marriage became turbulent and adulterous. There was a contentious divorce, and in 1916, Gladys and a brother shot Ed to death in front of their father's bank—and in front of the Sims daughters. (Courtesy Charles Anderson Sr.)

The death of Ed Sims triggered the last blood feud in Texas. Judge Cullen Higgins, one of the first attorneys in West Texas, skillfully secured the exoneration of both Gladys and her brother. The oldest son of Pink Higgins, Judge Higgins was brutally assassinated in 1918. Lawmen then halted two years of violence, and Gladys was happily married to famed Texas Ranger Frank Hamer, who also played a role in the shooting. (Courtesy Samantha Usnick.)

110

Eight
FROM RODEOS TO REEL COWBOYS

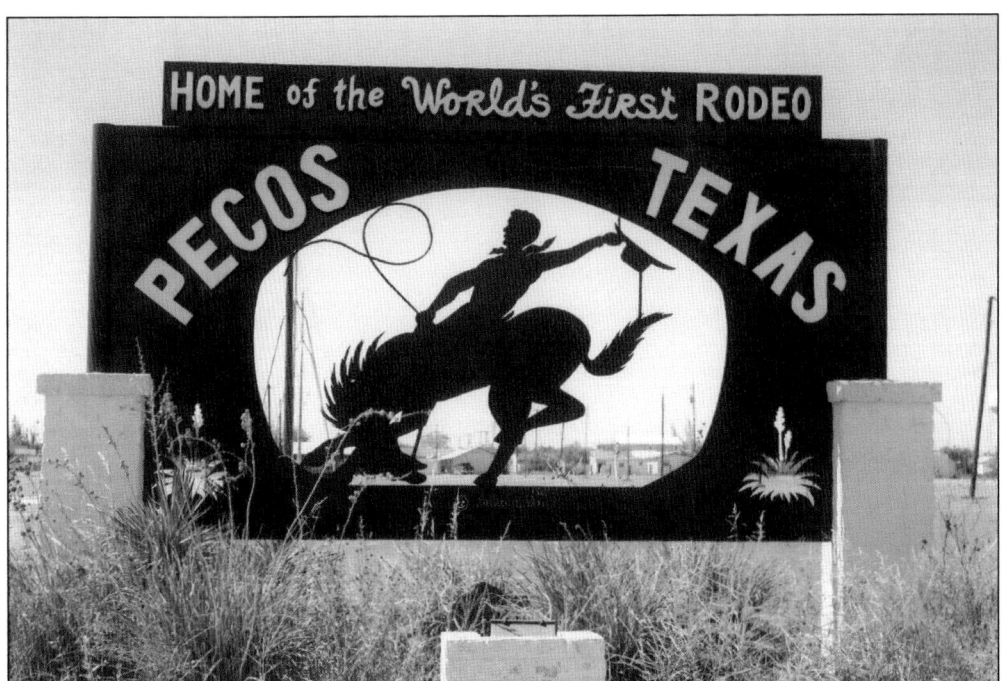

The town of Pecos proudly proclaims that the community is the "Home of the World's First Rodeo." Always a day of celebration in the West, the Fourth of July in 1883 was observed near the courthouse in Pecos with a calf-roping contest. This big sign is near the rodeo grounds on the east side of Pecos. (Photograph by the author.)

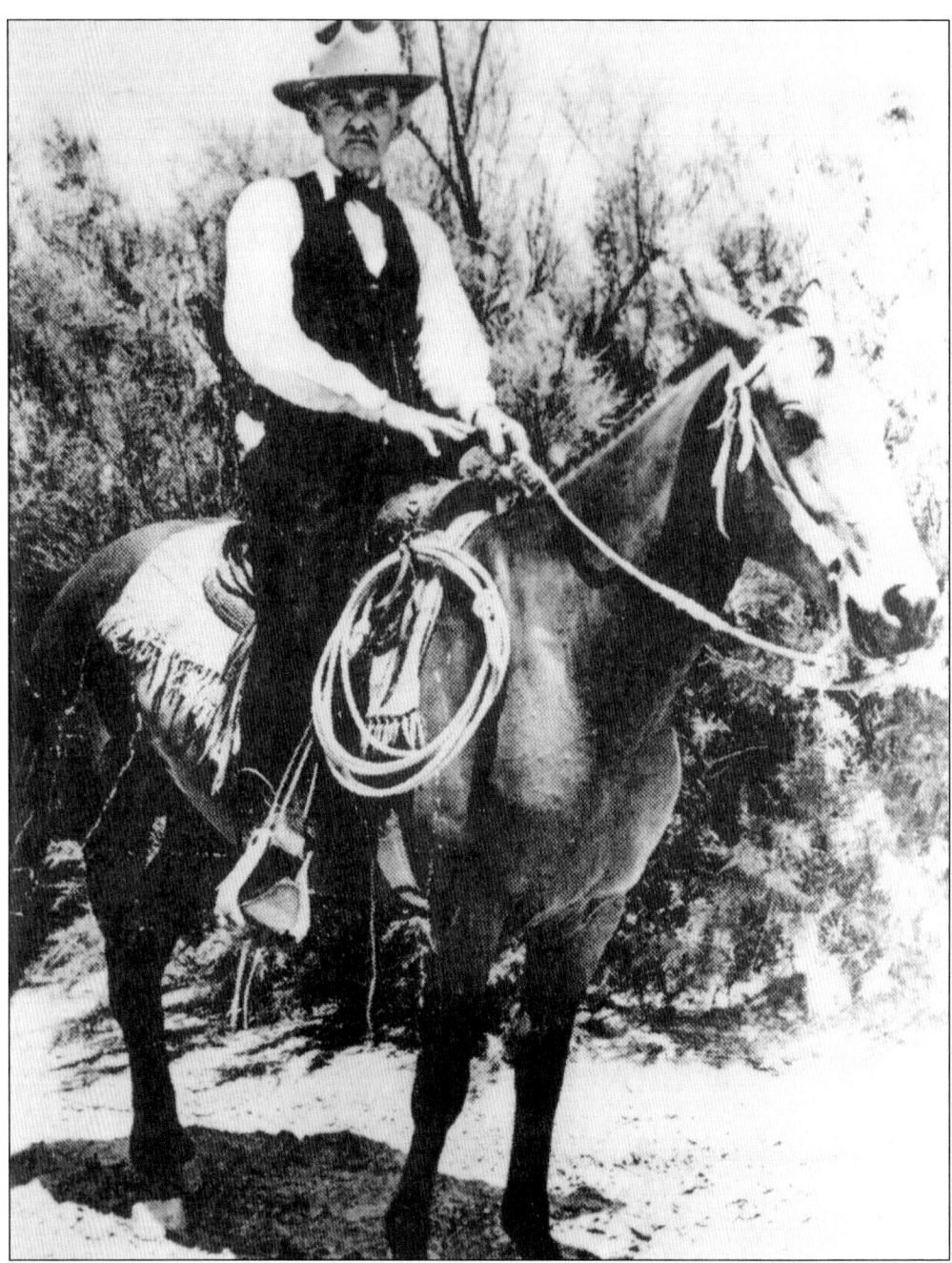

Cowboys from several area ranches each claimed that they had the fastest calf-ropers. It was decided to settle the matter when citizens gathered in Pecos City (population 300) to celebrate the Fourth of July with a picnic. On July 4, calves were driven to the Reeves County courthouse, while everyone in town watched. A calf-roping contest then ensued, with cowboys chasing calves down the main street of town. Although there were no cash prizes, winning cowboys were "treated" by the losers, and blue "ribbons" were cut by a pocket knife from the dress of a little girl. One of the participants, Henry Slack, lived in Pecos the rest of his life and led the rodeo parade every year. (Courtesy *Pecos, A History of the Pioneer West*.)

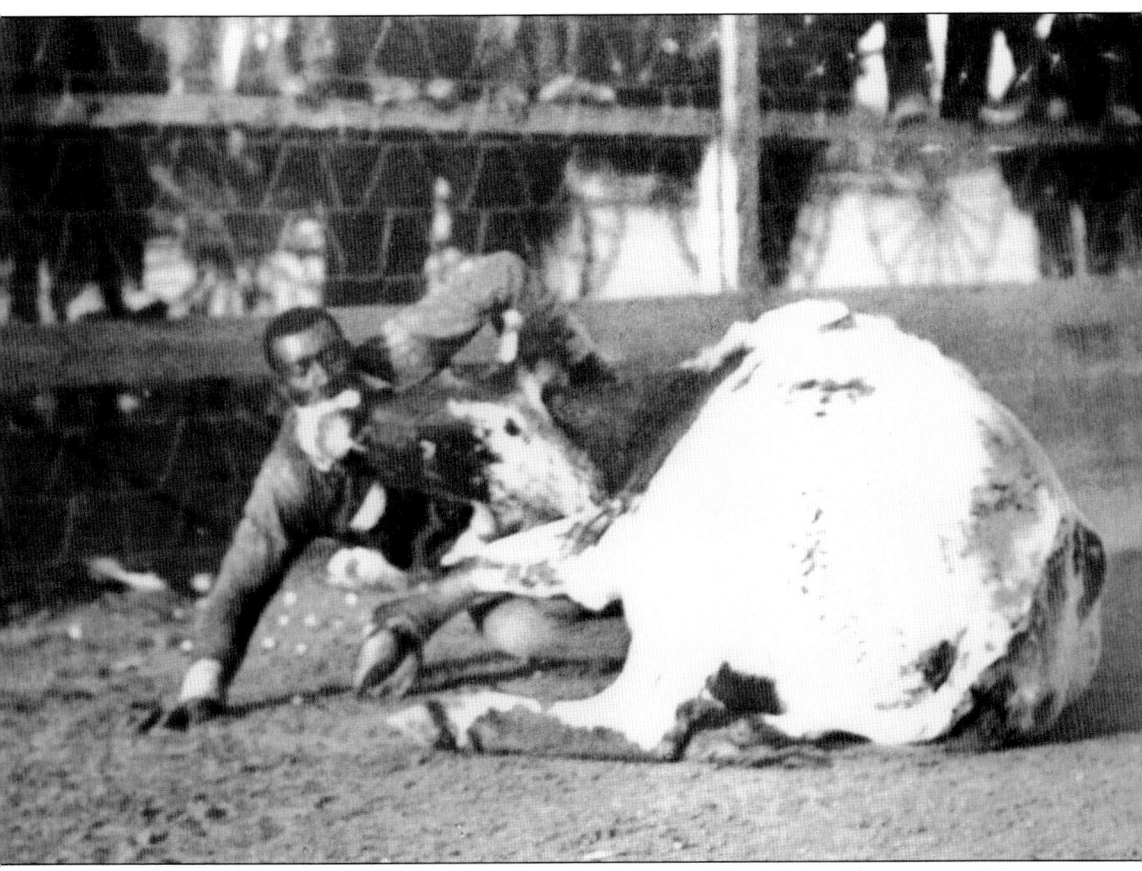

Growing up in Texas, young Willie Pickett observed cowboy life with rapt fascination. Many cattlemen used bulldogs to work their livestock. The dogs controlled steers by biting their lips, and one day Willie gave it a try. The boy seized a calf by the ears, sank his teeth into the animal's upper lip and, to the astonishment of cowboy onlookers, threw it with ease. "You took the bulldog's job away from him when you bulldogged that steer!" exclaimed one cowboy. Later, while cowboying in brush country where it was hard to build a loop, Pickett would lean from his horse, grip the steer's long horns, and successfully bulldog it. When he became a popular rodeo and Wild West Show performer, "Bulldoggin' Bill" Pickett lost several teeth in the line of duty. Pickett single-handedly created the rodeo event known as bulldogging and eventually became the first black man elected to the National Rodeo Hall of Fame. (Author's collection.)

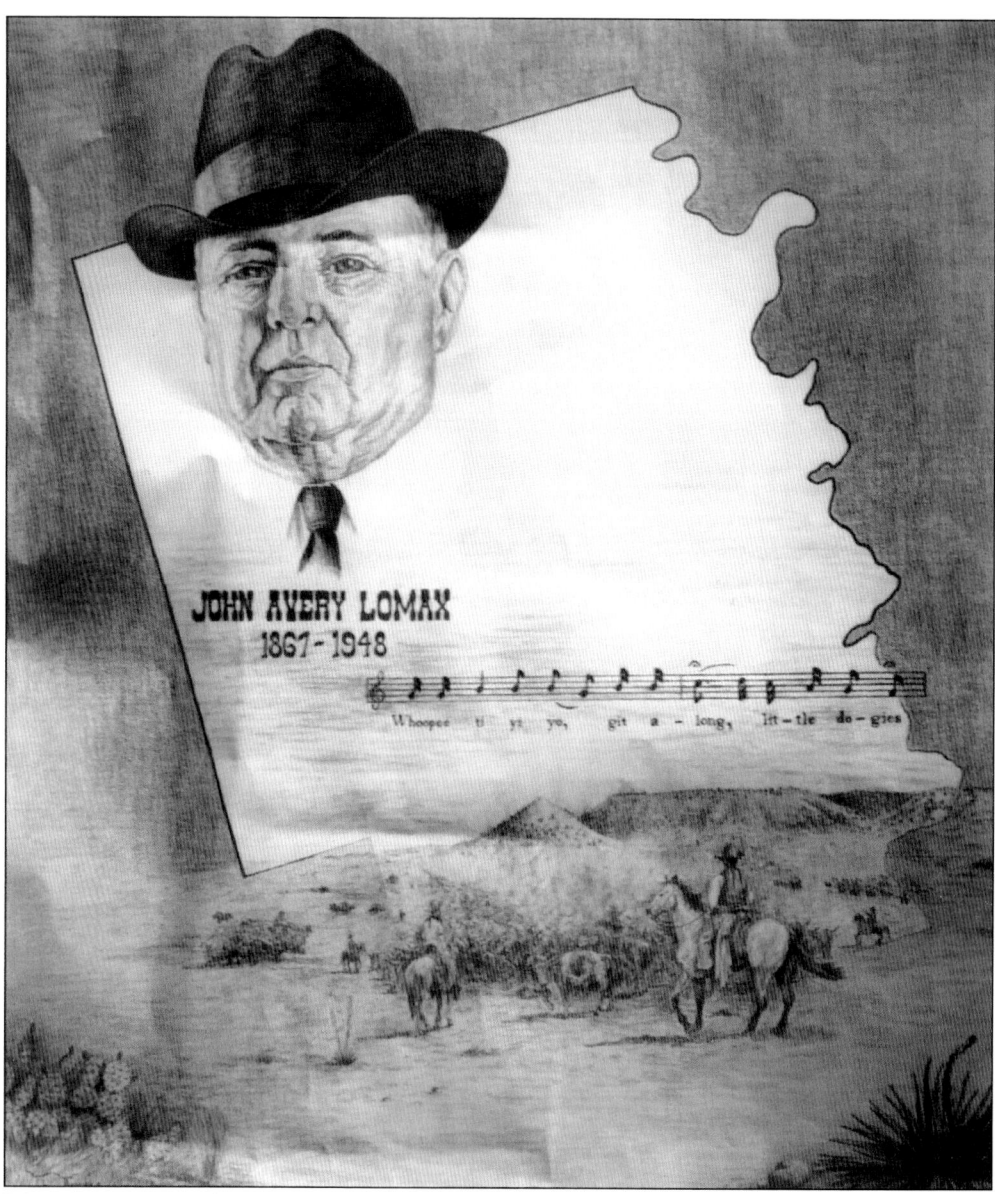

John A. Lomax was born in Mississippi in 1867, but two years later, his family moved to Bosque County, Texas. The Lomax farm stood beside a branch of the Chisholm Trail, and John was captivated by the cowboy ballads he heard. By the 1880s, he began to write down these folk songs. Lomax attended the University of Texas, married, taught at Texas A&M, and in 1906, received a scholarship to Harvard. His professors urged him to intensify his collecting. In Fort Worth, San Antonio, Abilene, and other Texas towns, he learned versions of "The Old Chisholm Trail," "Git Along, Little Dogies," "Home on the Range," and other cowboy ballads. His first collection, *Cowboy Songs and Other Frontier Ballads*, was published in 1910, and his lifelong search saw him travel 200,000 miles. (Courtesy Bosque County Collections, Meridian.)

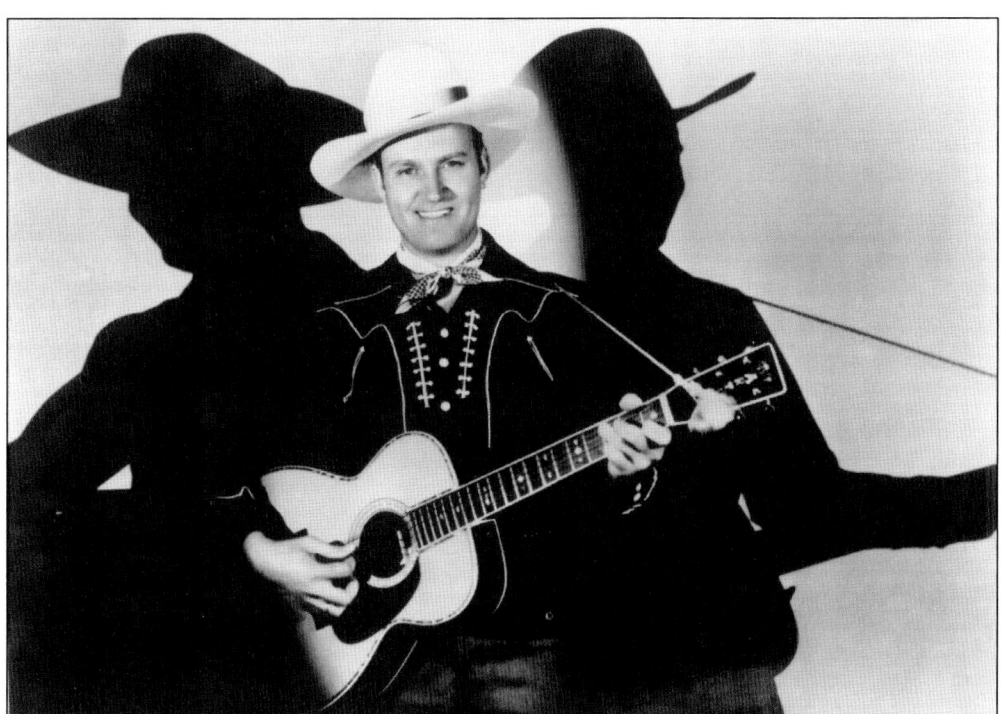

As the first "singing cowboy" of the silver screen, Gene Autry (above), of Tioga, Texas, made an enormous contribution to popularizing the music of the range. Successful as a radio and recording performer, Autry came to Hollywood in 1934 and soon achieved stardom. He would film more than 90 "horse operas," and at the height of his popularity, he received 80,000 fan letters a week. (Author's collection.)

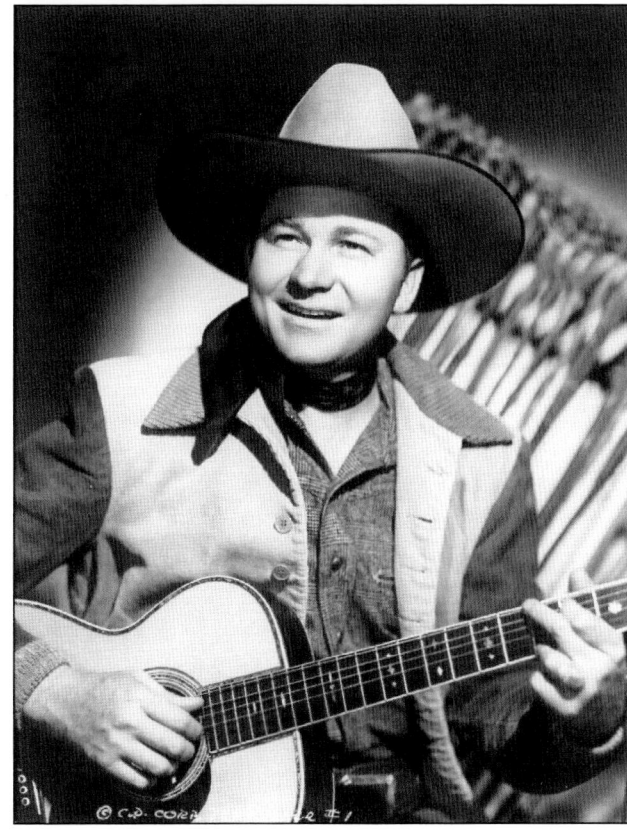

Tex Ritter (right) was influenced at the University of Texas by John A. Lomax and J. Frank Dobie. He brought his collection of cowboy ballads to New York, where he spent nearly a decade singing and acting onstage and on network radio. Brought to Hollywood in 1936, he starred in 60 Western movies and enjoyed a long recording career. (Author's collection.)

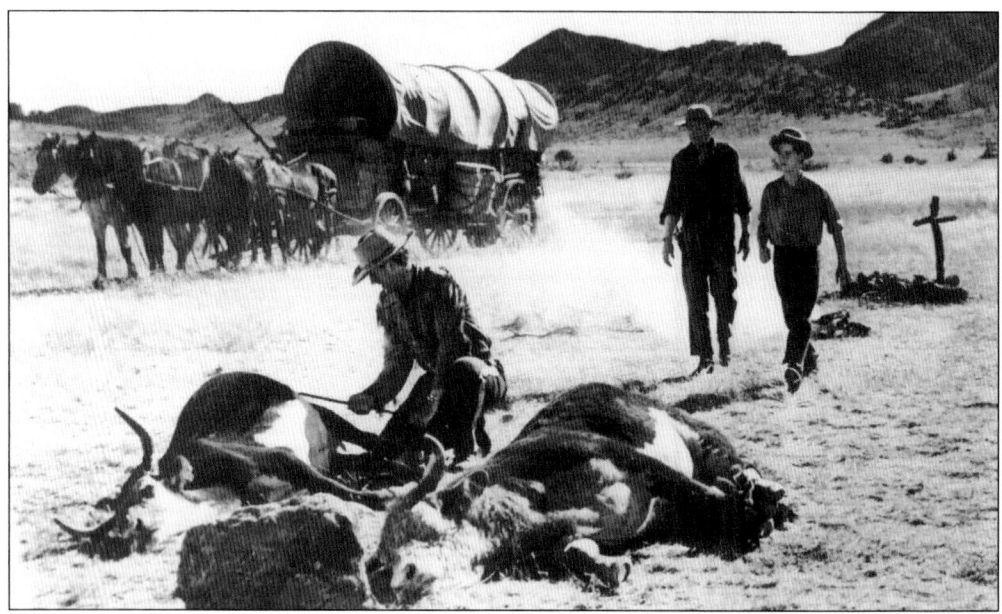

Filmed in 1948 and starring Hollywood's greatest Western icon, John Wayne, *Red River* is the ultimate trail drive movie. Early in the film, Wayne, assisted by costars Walter Brennan and Montgomery Clift, builds a Texas ranching empire. With 6,000 cattle onscreen and all the power of an epic motion picture, the rest of the movie is a dramatic, adventurous trail drive to Abilene, Kansas. (Author's collection.)

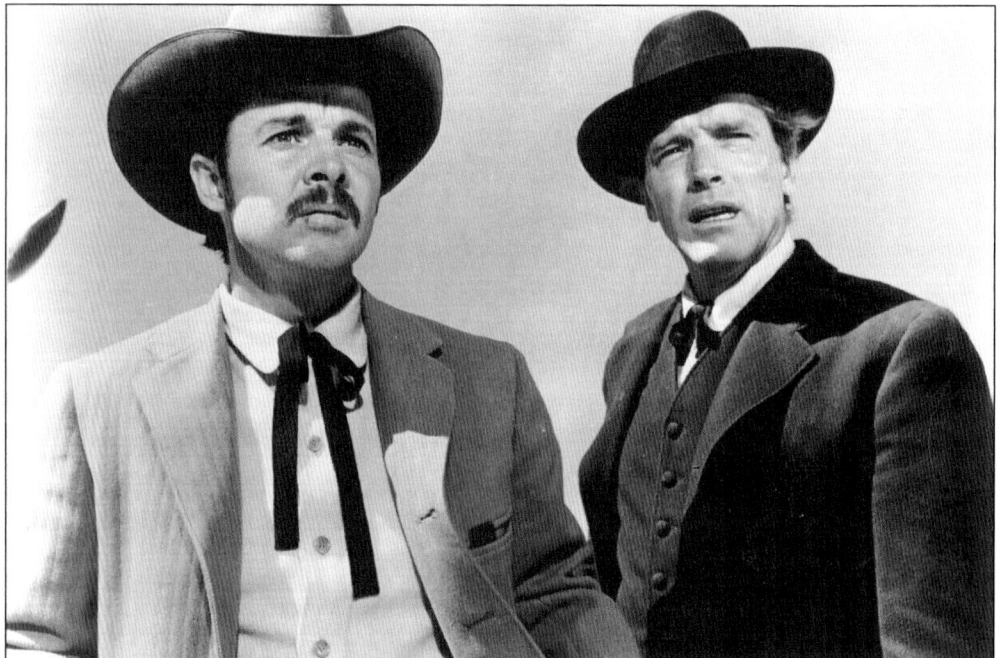

In 1960, famed director John Huston released *The Unforgiven*, based on a superb novel of the Texas cattle frontier written by Alan LeMay, who also penned another classic, *The Searchers*. Burt Lancaster (right) and Texas war hero Audie Murphy (left), in one of his finest performances, engage in a vicious struggle against Kiowa warriors from their isolated ranch. (Author's collection.)

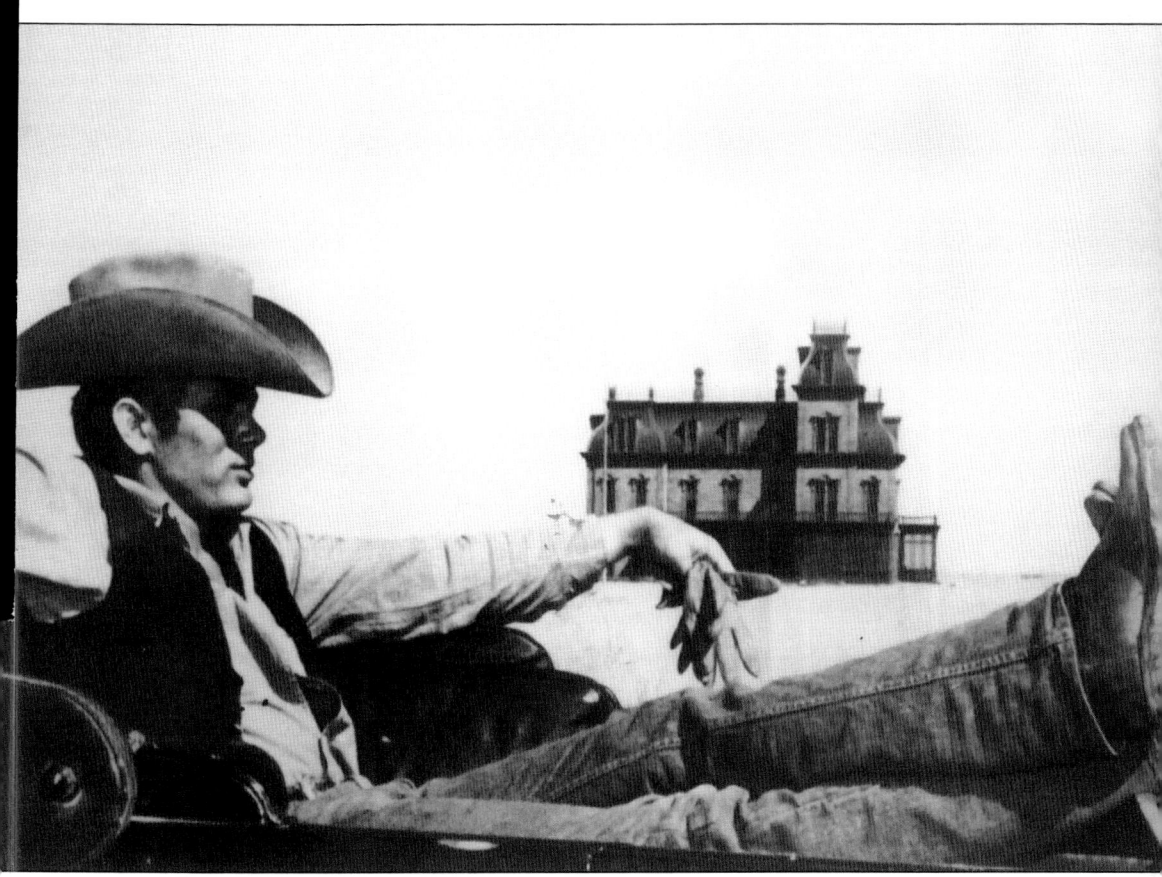

Giant, Edna Ferber's novel based loosely on the King Ranch, was adapted for the screen with stars Rock Hudson, Elizabeth Taylor, James Dean, Chill Wills, and Monte Hale, of San Angelo. The 1956 motion picture was filmed at Marfa, where the picturesque El Paisano Hotel served as crew headquarters, and the local movie theater was used as the nightly screening room by director George Stevens. In the film, Tejano vaqueros provide the workforce for Hudson's vast Reata Ranch, and Elizabeth Taylor labors against racial injustice. James Dean (above) steals scene after scene as a Reata ranch hand turned oil tycoon. There are superb cowboy scenes, as well as barbeques and fistfights and a magnificent musical score. Giant was long called "The National Movie of Texas." (Author's collection.)

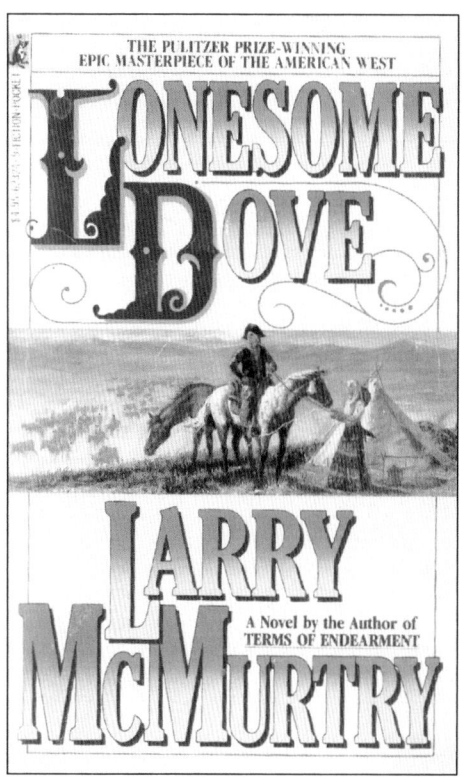

Larry McMurtry was raised on a cattle ranch near Archer City. A gifted writer, he enjoyed early success with the novella *Horseman, Pass By*, which was made into the popular film *Hud*. Starring Paul Newman as an amoral rancher, the 1963 motion picture was filmed in northwest Texas and won three Academy Awards. In 1985, McMurtry's towering cattle drive novel, *Lonesome Dove*, became a bestseller and won the Pulitzer Prize. (Photograph by the author.)

The genesis of *Lonesome Dove* was a 1970s screenplay by McMurtry that would have starred John Wayne, James Stewart, and Henry Fonda. When Wayne declined, however, the project was dropped. But years later, McMurtry's subsequent novel was turned into a memorable miniseries starring Robert Duvall and Texan Tommy Lee Jones, who still operates a ranch near San Saba. The climax of *Lonesome Dove* is a version of the death of Oliver Loving (Duvall, left) and the return of his corpse to Texas by Goodnight (Jones, right). (Author's collection.)

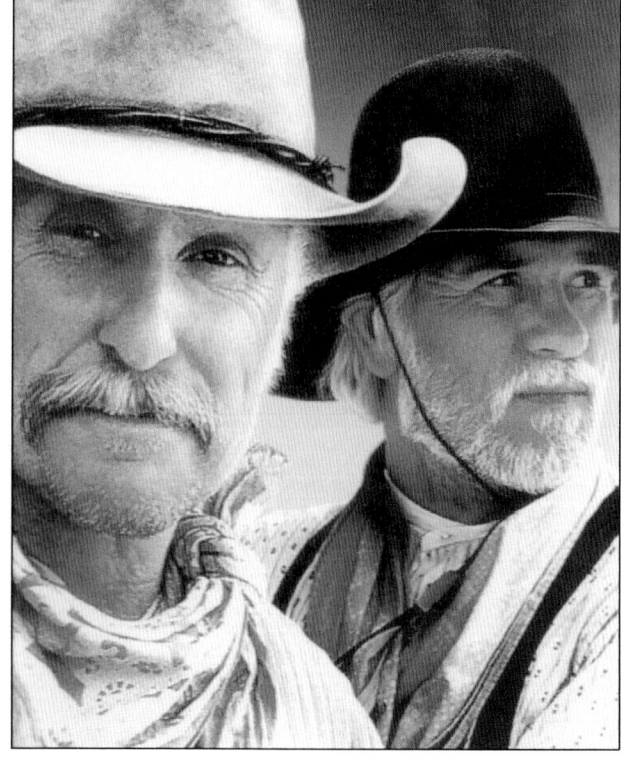

Nine
WHERE TO GO AND WHAT TO SEE

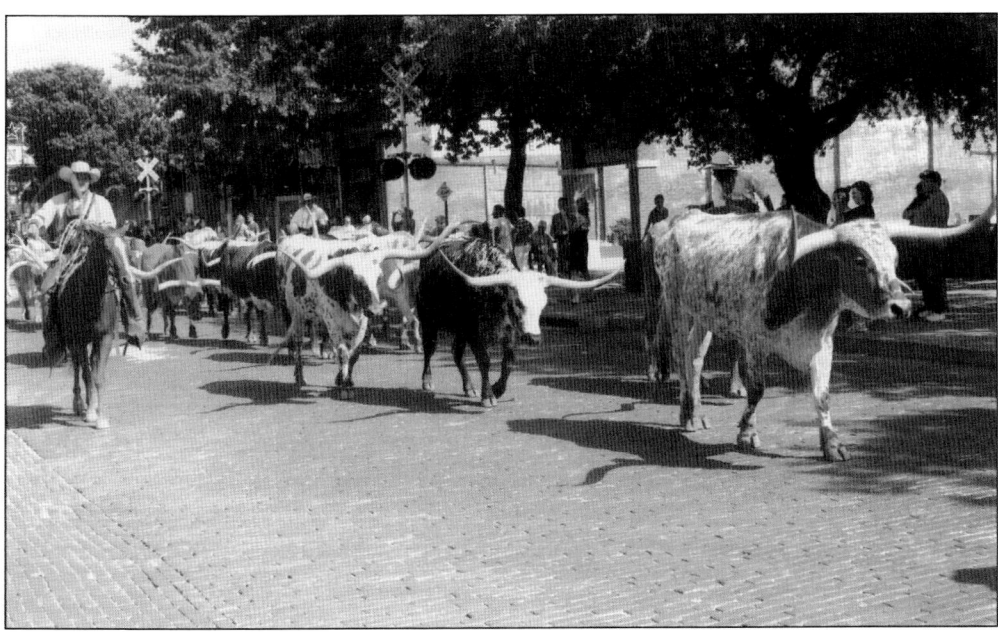

Twice each day, at 11:30 a.m. and 4:00 p.m., a herd of longhorns is driven along Exchange Avenue at the Fort Worth Stockyards. Spectators watch cowboys in authentic dress drive cattle with incredible sets of horns. The feel of the Old West is further recaptured throughout the entire Stockyards District. (Photograph by the author.)

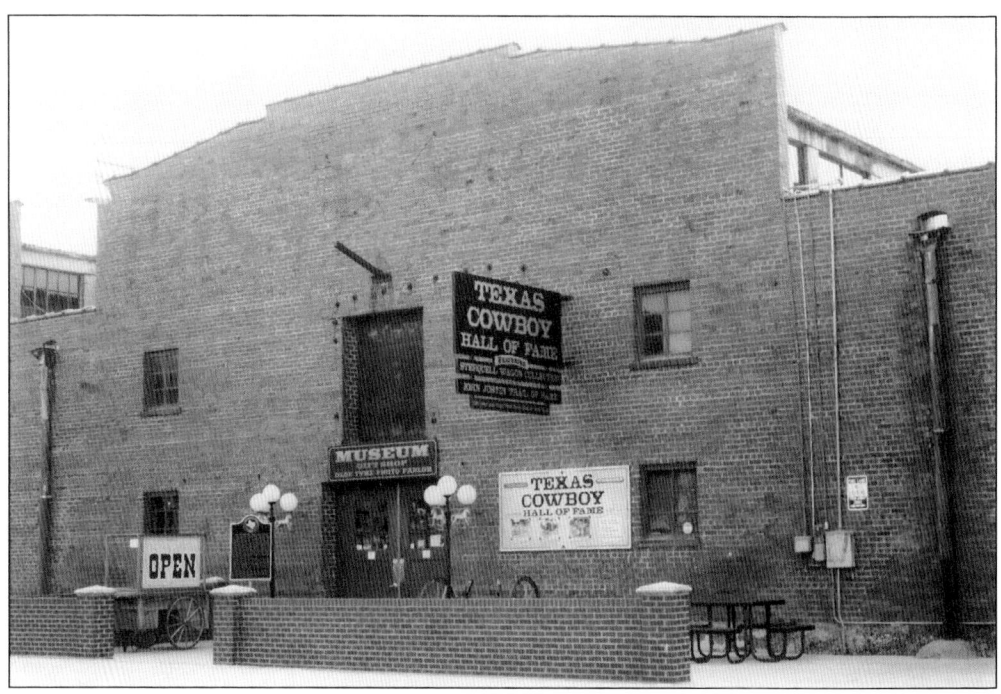

One of the finest attractions of the Stockyards District is the diverse array of cowboy equipment and clothing displayed in the Texas Cowboy Hall of Fame. (Photograph by the author.)

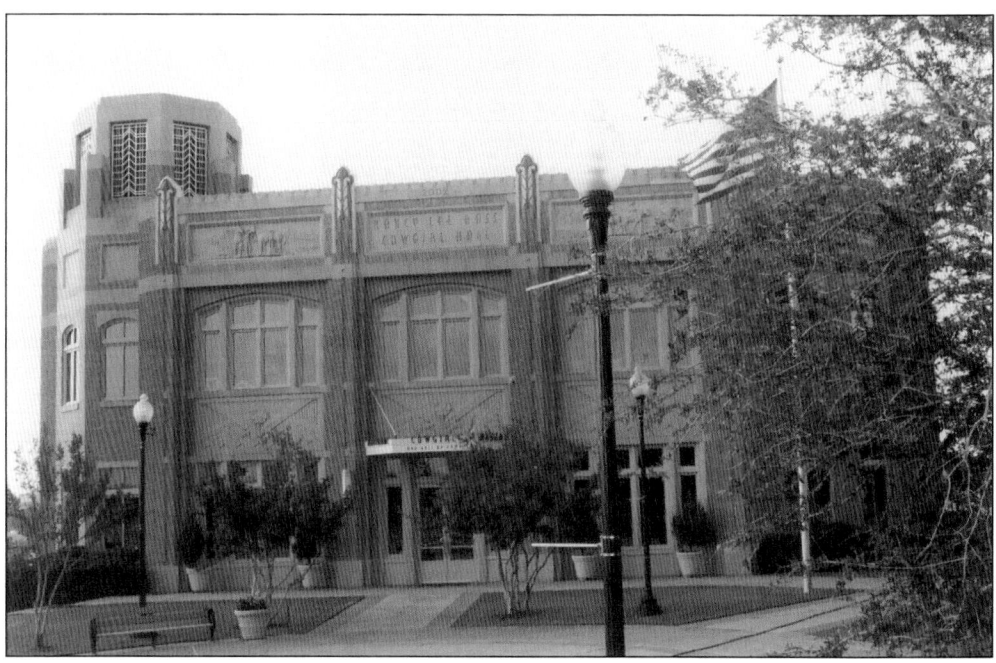

Fort Worth's National Cowgirl Museum and Hall of Fame documents and honors the women of the American West, from ranch wives and cowgirls to Wild West Show and rodeo performers, Western movie actresses, and country singers. This state-of-the-art facility is a delight for folks of all ages. (Photograph by the author.)

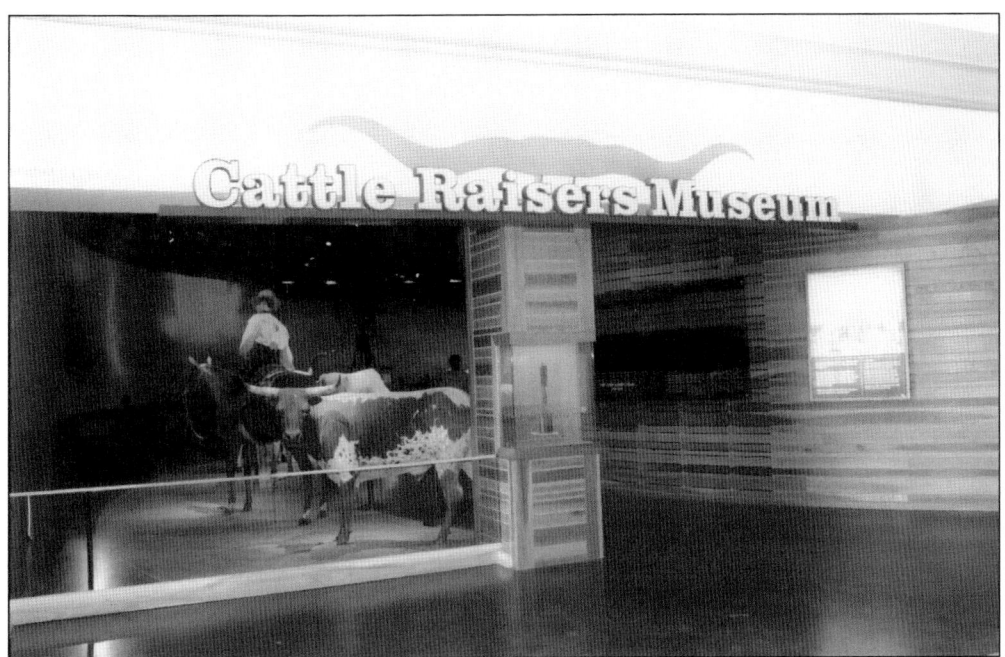

The Southwestern Cattle Raisers Association Museum is part of the Fort Worth Museum of Science and History. The museum is packed with artifacts and information from the early cattle industry. (Photograph by the author.)

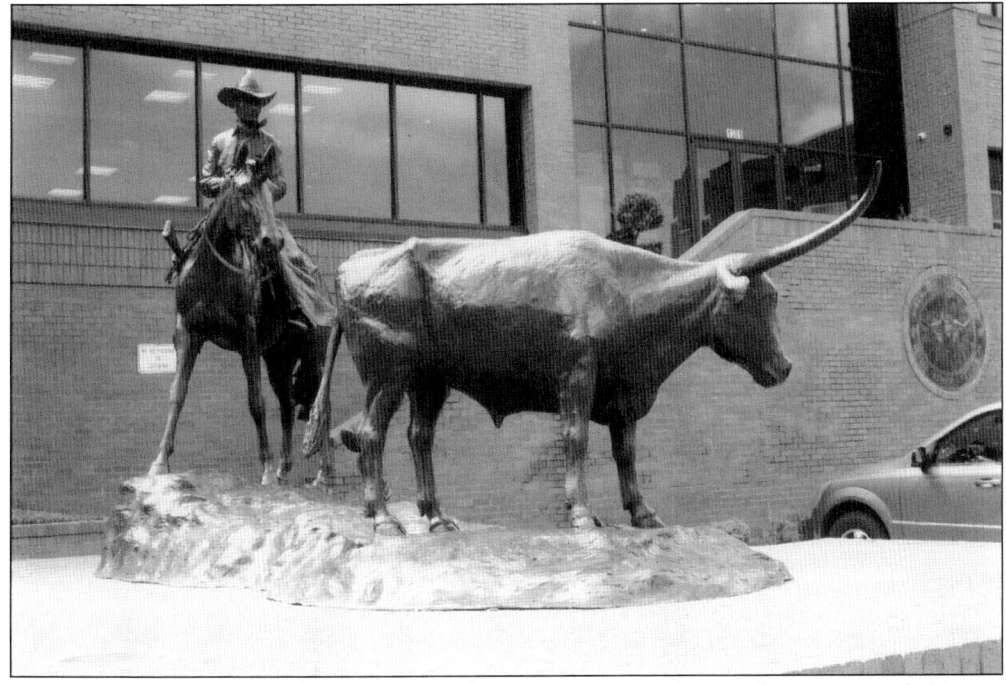

The SWCRA Museum was long housed at the headquarters of the Southwestern Cattle Raisers Association near downtown Fort Worth. Superb cattlemen's statuary, as well as a historical branding wall, can still be viewed at SWCRA headquarters. (Photograph by the author.)

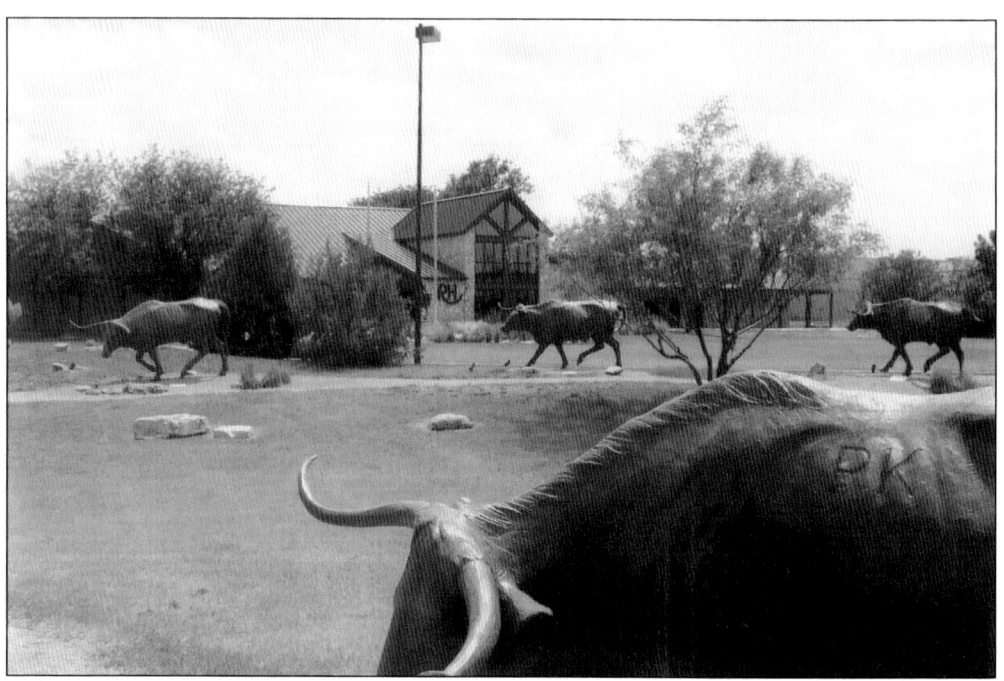

The internationally famous National Ranching Heritage Center in Lubbock is housed in a recently expanded main building with superb range statuary dominating the approach. Inside are fascinating exhibits of artifacts and artwork. (Photograph by the author.)

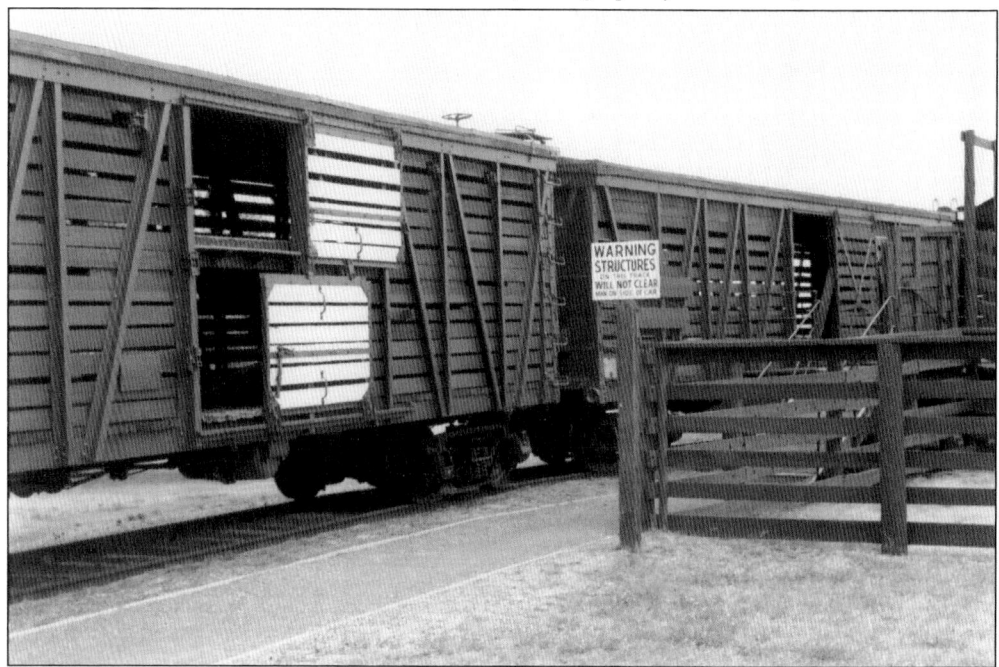

On a 16-acre site behind the National Ranching Heritage Center's main building, the story of ranching is told through an incomparable collection of nearly 50 historical structures. There are dugouts, an XIT bunkhouse, a 6666 barn, ranch houses of varied sizes, and auxiliary buildings of every type, all authentically furnished and equipped. (Photograph by the author.)

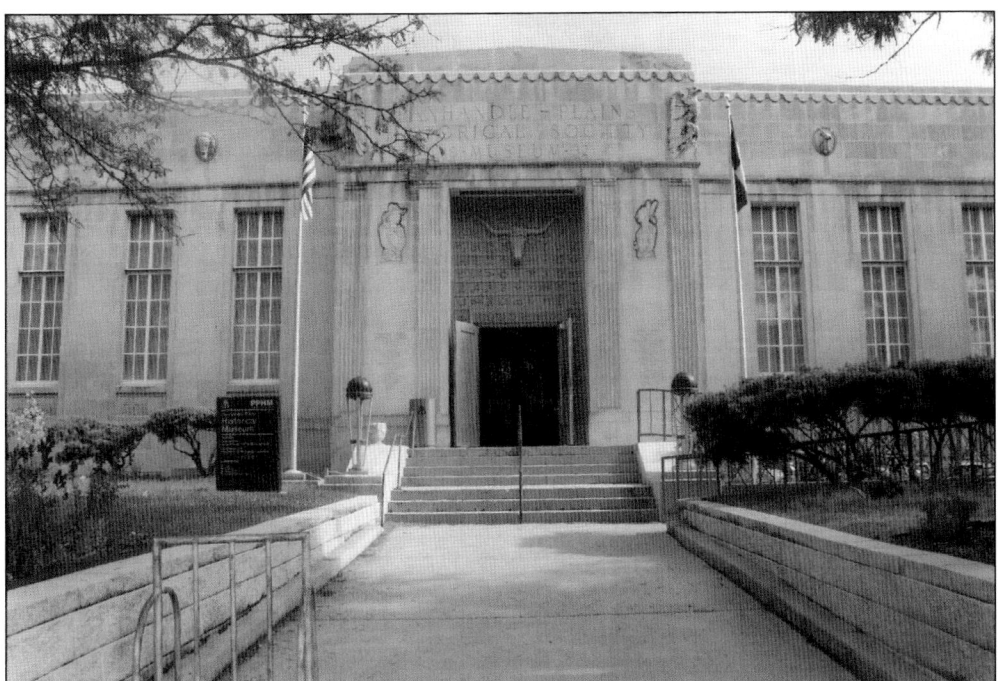

The Panhandle-Plains Museum in Canyon (above) is the oldest museum in Texas. The holdings are enormous and varied, but there is a great deal of focus on cowboys and ranching. Outside is the historic T-Anchor Ranch headquarters, as well as an impressive statue of Charles Goodnight. (Photograph by the author.)

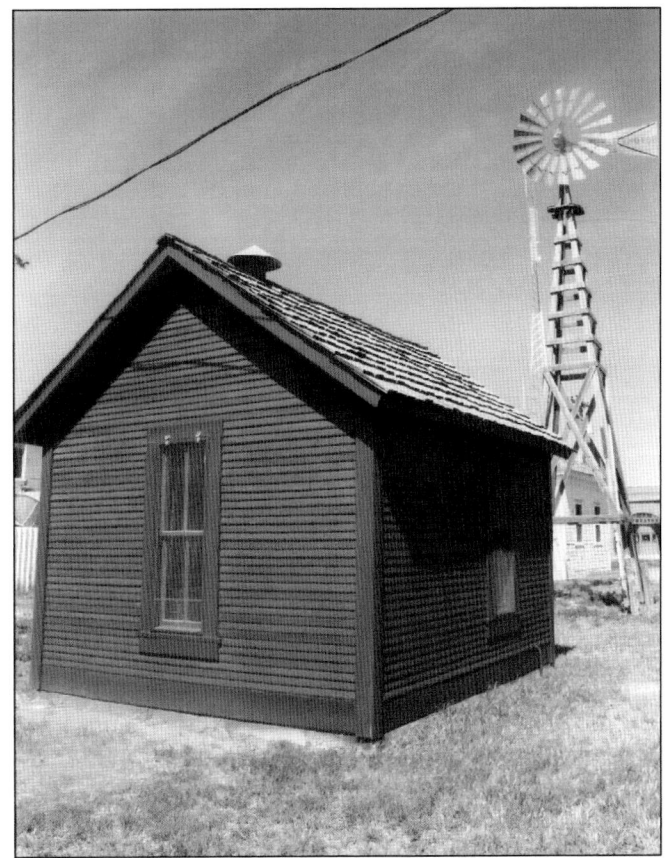

This line shack and windmill have been brought into Claude for display at the Armstrong County Museum. (Photograph by the author.)

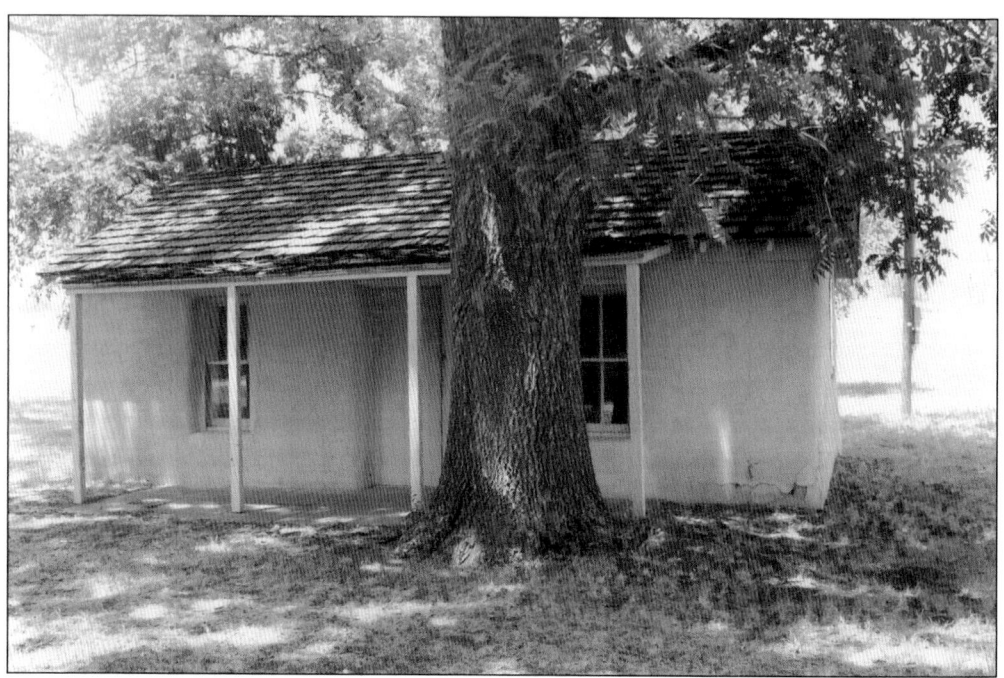

At Doan's Crossing, on the Western Cattle Trail just south of the Red River, the adobe house of Judge Corwin F. Doan and his wife, Lide, has been preserved. The Doans frequently entertained cattlemen in their home. (Photograph by the author.)

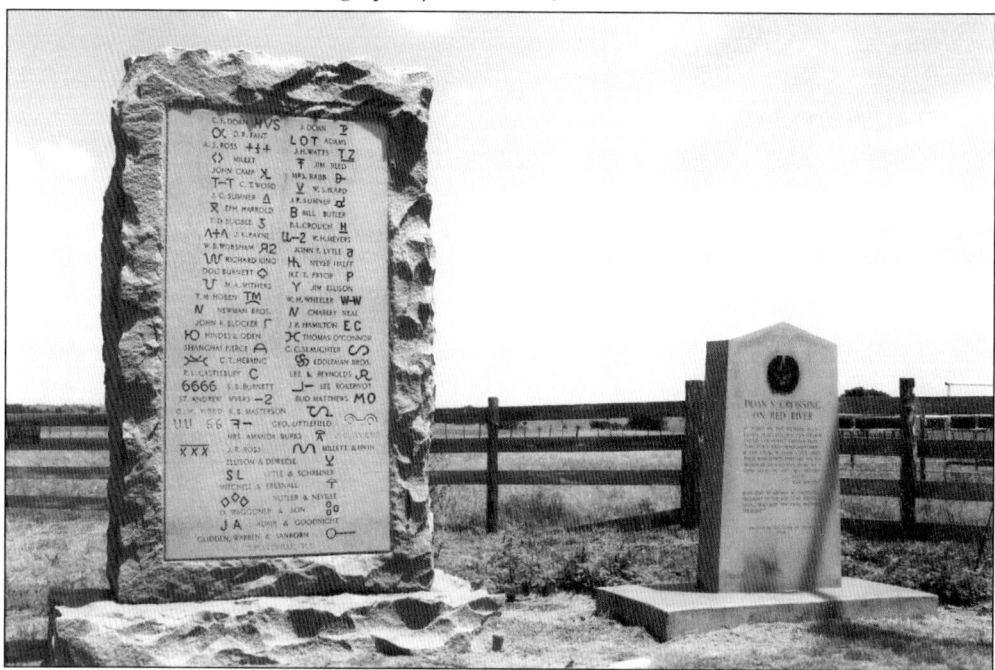

One of the monuments marking the approach to Doan's Crossing on the Red River shows famous brands of the millions of cattle that crossed at Doan's. Among the noted cattlemen listed on this monument are Charles Goodnight, C.C. Slaughter, Richard King, Shanghai Pierce, Burke Burnett, George Littlefield, D.R. Fant, John Lytle, and Dan Waggoner. (Photograph by the author.)

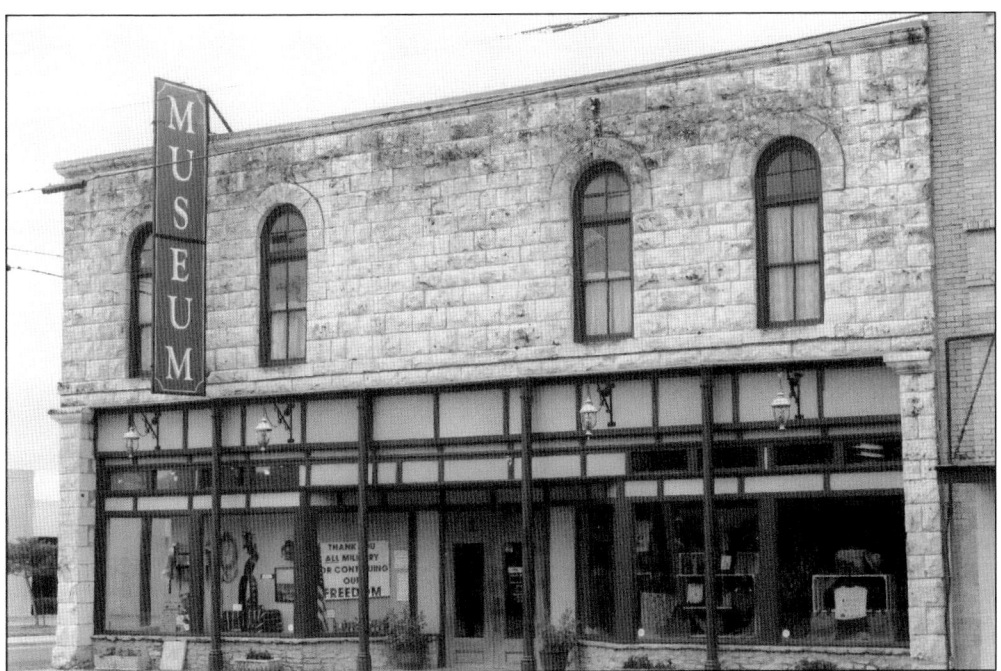

Gatesville's Coryell County Museum and Historical Museum features the world's largest spur collection, along with associated cowboy displays. Among many other treasures, the first county jail, of double timber construction, is on display. (Photograph by the author.)

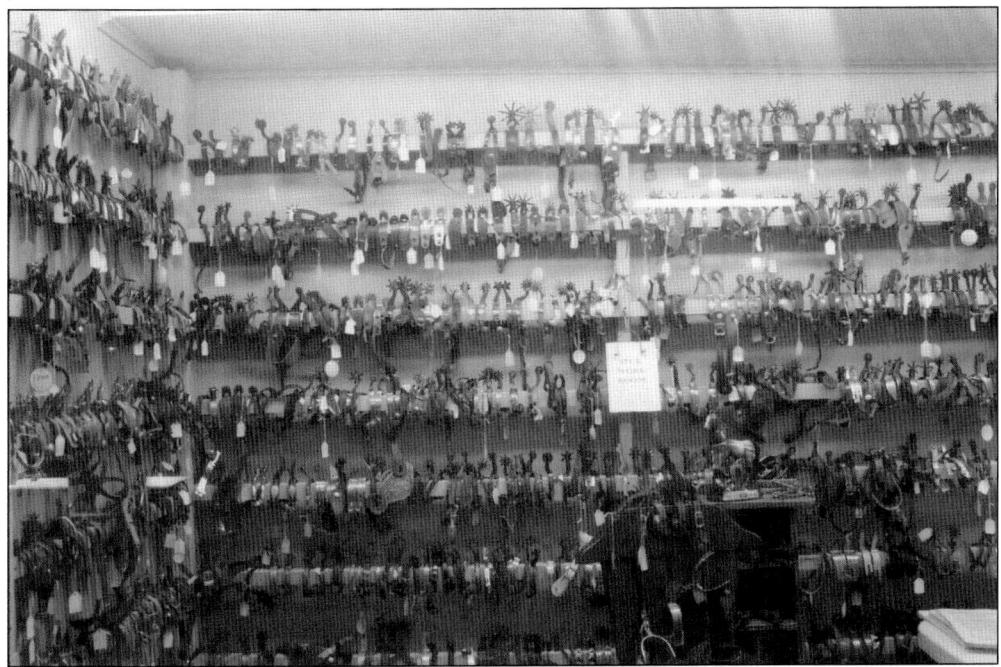

The Lloyd and Madge Mitchell Collection, seen here, is comprised of more than 7,000 pairs of spurs. Lloyd Mitchell was a longtime high school football coach who became a passionate spur collector. Every type of spur is found in the displays, including elegant spurs belonging to famous personages. (Photograph by the author.)

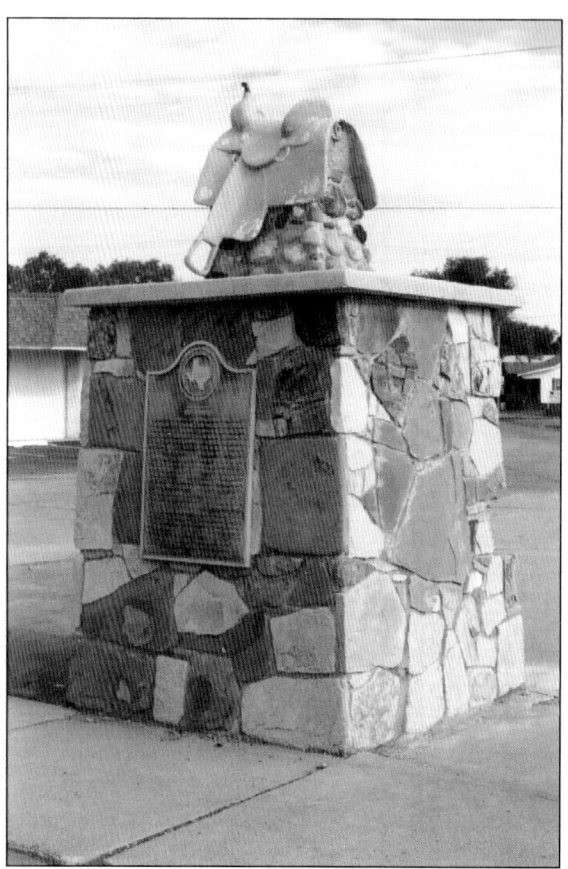

The brick headquarters building of the enormous XIT Ranch, a structure of great historical importance to ranching history, is in Channing. To the north of Channing, facing the courthouse in Dalhart, is the XIT Historical Museum, which is crammed with artifacts and photographs of the famous ranch. The Empty Saddle Monument (left) was built in 1940 on Highway 87 to honor all XIT cowboys. (Photograph by the author.)

The Waco Suspension Bridge (below) was built across the Brazos River in the late 1860s as a toll bridge for northbound cattle herds and other traffic. Today, it is maintained as a pedestrian bridge, with a magnificent sculpture of a Texas cowboy driving longhorns onto the bridge. (Photograph by the author.)

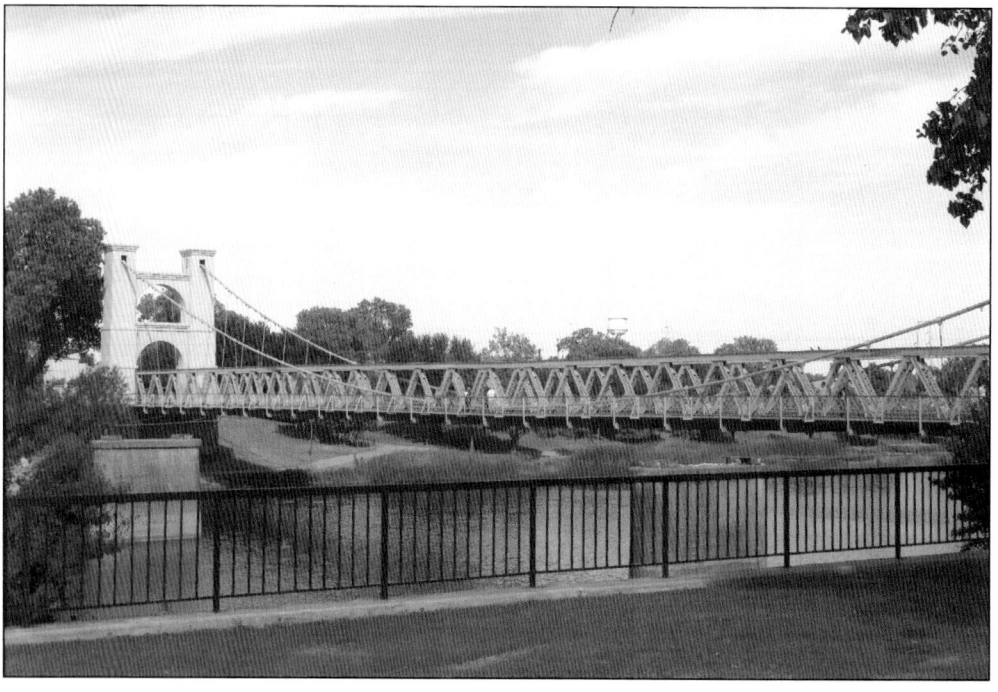

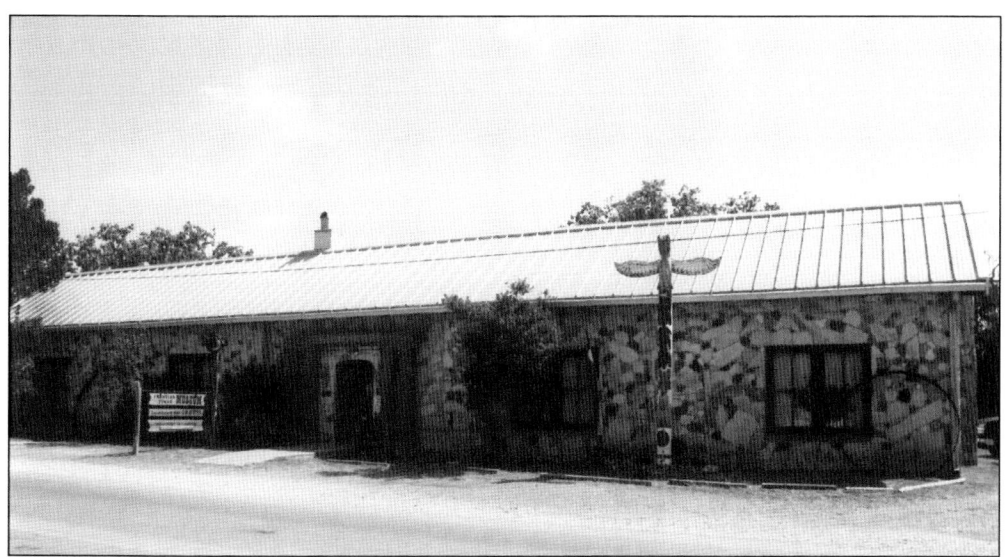

The Frontier Times Museum in Bandera was established in 1923 by J. Marvin Hunter Sr. A journalist, historian, editor, and author, Hunter built the museum to exhibit his growing collection of Western artifacts. He founded *Frontier Times*, a magazine devoted to pioneer history. Later, he also published *True West* and *Old West* magazines. (Photograph by the author.)

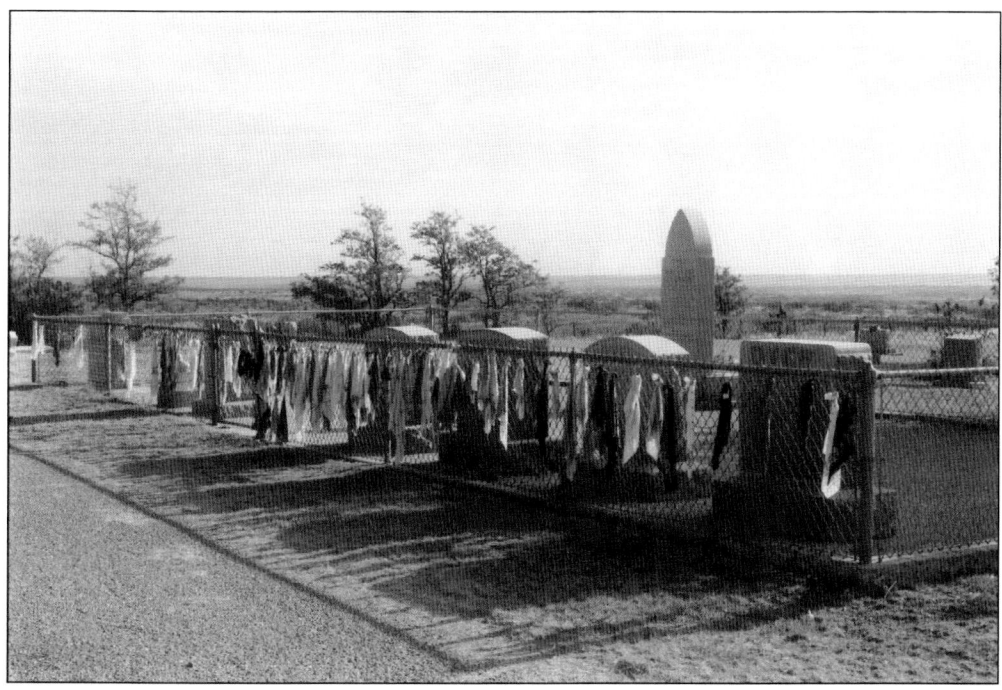

In the village of Goodnight, Charles Goodnight's magnificently restored home stands just south of Highway 287 (see page 40). Across the highway at the Goodnight Cemetery, the legendary cattleman is buried in a family plot. Visitors tie bandanas on the metal fence, but after a few months, the weathered cloths are removed. Soon, however, new bandanas always appear on the fence. (Photograph by the author.)

Discover Thousands of Local History Books Featuring Millions of Vintage Images

Arcadia Publishing, the leading local history publisher in the United States, is committed to making history accessible and meaningful through publishing books that celebrate and preserve the heritage of America's people and places.

Find more books like this at
www.arcadiapublishing.com

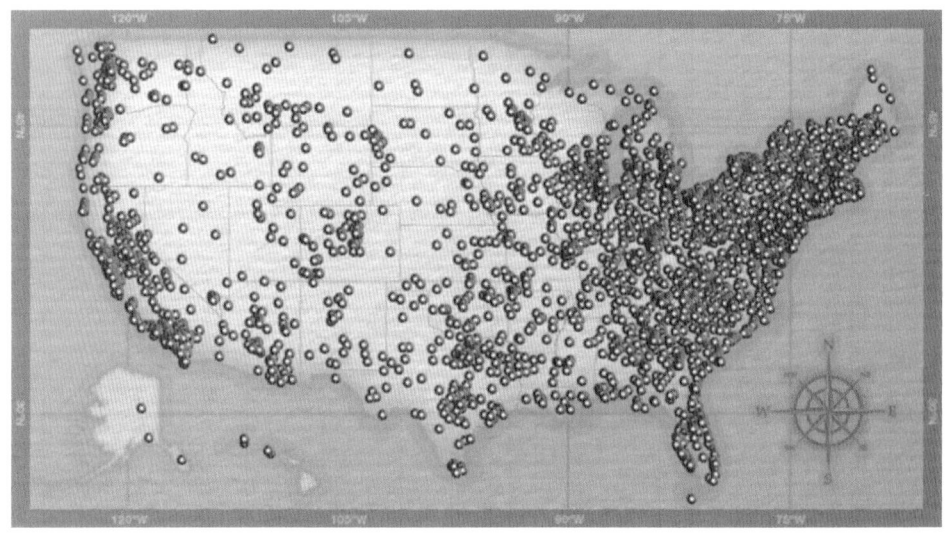

Search for your hometown history, your old stomping grounds, and even your favorite sports team.

Consistent with our mission to preserve history on a local level, this book was printed in South Carolina on American-made paper and manufactured entirely in the United States. Products carrying the accredited Forest Stewardship Council (FSC) label are printed on 100 percent FSC-certified paper.

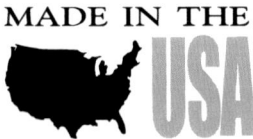